Automotive Systems Engineering II

Hermann Winner • Günther Prokop •
Markus Maurer
Editors

Automotive Systems Engineering II

Springer

Editors
Hermann Winner
Fachgebiet Fahrzeugtechnik
Technische Universität Darmstadt
Darmstadt, Germany

Günther Prokop
Institut für Automobiltechnik
Technische Universität Dresden
Dresden, Germany

Markus Maurer
Institut für Regelungstechnisch
Technische Universität Braunschweig
Braunschweig, Germany

ISBN 978-3-319-61605-6 ISBN 978-3-319-61607-0 (eBook)
DOI 10.1007/978-3-319-61607-0

Library of Congress Control Number: 2013935997

Printed on acid-free paper

This Springer imprint is published by Springer Nature
The registered company is Springer International Publishing AG
The registered company address is: Gewerbestrasse 11, 6330 Cham, Switzerland

Preface

Automotive Systems Engineering (ASE) addresses cross-functional and interdisciplinary aspects of systems engineering for road vehicles. Some of the approaches originate from the systems engineering “world” of different product categories; others are very specific to the automotive world, especially when the addressed problem first became evident there.

The challenge of functional safety does not have its origin in automotive applications, but since the last two decades, it has revolutionized the processes of how we develop automotive products. Starting with top-down oriented system architectures, systematic development of functions and validation by a suitable qualification process are the key factors for successful control of complexity.

With the progress of technologies in environmental perception and cognition, the automotive world is now pioneering the challenge of autonomous acting in a public space. Autonomous driving substitutes tasks from a human and shifts them to a robot. As we know from the high number of road traffic accidents and their consequences, driving always contains a high potential risk. Methods to minimize the risk and to ensure the safety of autonomous driving are in the foreseeable future but not achieved yet.

The change to ASE is not limited to future products. The development process of traditional automobiles needs improvements due to the immense effort and costs for supporting the growing variety of models. Two examples for the rethinking of the process are shown in this edition. One is the design of ride comfort characteristics on a subsystem level during the product development process. The other shows methods for change management in automotive release processes.

The chapters of the volume reflect the work of just few institutes and cannot represent the whole variety of ASE. However, we think it representatively shows the width and depth of modern research approaches for that field.

We wish our readers stimulating reading and look forward to receiving a wide spectrum of feedback.

Darmstadt, Germany Hermann Winner
Dresden, Germany Günther Prokop
Braunschweig, Germany Markus Maurer

Contents

Part I
Development Process

Chapter 1
Design of Ride Comfort Characteristics on Subsystem Level in the Product Development Process

Christian Angrick, Günther Prokop, and Peter Knauer

Abstract In the automotive development process the significance of full vehicle ride comfort is becoming more important. Due to rising complexity and new boundary conditions upcoming in the development process, like a higher variety of models, higher functional demands, and decreasing development times, the design of respective ride comfort characteristics in early phases of the development is desirable. The necessity for a precisely defined and structured process is therefore increasing. In driving dynamics already a high progress is achieved in defining a respective process, which can be essentially attributed to the application of a subsystem level in the derivation of vehicle properties. In ride comfort however, the progress is less advanced, as no comparable subsystem methods or models exist. Therefore in the following the focus lies specifically on the integration of a subsystem level in the derivation process of vehicle properties from full vehicle to components. For that purpose, initially the automotive development process will be illustrated in its general structure and its specific realization in driving dynamics and ride comfort. The advantages and disadvantages of the respective disciplines will be emphasized. Furthermore the structure of subsystem models in ride comfort as well as associated concept parameters are introduced. In consideration of the new methodology, the integration within the automotive development process is illustrated and examples are given. Finally the findings of the investigation are summarized and the advantages of the methodology are emphasized.

C. Angrick (✉)
AUDI AG, I/EF-13, 85045 Ingolstadt, Germany

TU Dresden, Institut für Automobiltechnik Dresden - IAD, Lehrstuhl für Kraftfahrzeugtechnik, George-Bähr-Straße 1c, 01062 Dresden, Germany
e-mail: christian.angrick@audi.de

G. Prokop
TU Dresden, Institut für Automobiltechnik Dresden - IAD, Lehrstuhl für Kraftfahrzeugtechnik, George-Bähr-Straße 1c, 01062 Dresden, Germany

P. Knauer
AUDI AG, I/EF-13, 85045 Ingolstadt, Germany

H. Winner et al. (eds.), *Automotive Systems Engineering II*,
DOI 10.1007/978-3-319-61607-0_1

Keywords Automotive • Ride comfort • Subsystem • Development process • Simulation • Target cascading • Derivation process • Concept model • Evaluation • Driving dynamics

1.1 Introduction and Objective Targets

With rising complexity and new boundary conditions upcoming in the development process of vehicles,[1] like a higher variety of models, higher functional demands, and decreasing development times (Rauh 2003, p. 135), it is necessary to specify processes which allow for a structured derivation of properties on different levels of detail of the vehicle. These are basically given by full vehicle, subsystem and component level, which can furthermore be divided in other meta levels. With respect to an initial level, the corresponding derivation of properties, also called target cascading, describes the process of determining adequate properties on sub levels, while the level of detail is continuously rising.

On full vehicle level characteristic values and targets for the respective discipline (e.g. driving dynamics and ride comfort) are defined. In the following, on subsystem level concept independent abstract parameters for characterizing the behavior of subassemblies are used. These are given for example by roll center height or toe compliance of a suspension, which can be described by characteristic scalar values or curves. On this level, the full vehicle is therefore described by a black box, without further knowledge of the individual concept of a subassembly. Finally, component properties are defined on the most detailed level. Exemplary, this can be bushing stiffnesses of an axle or the relaxation length of a tire. Overall, the target cascading aims at deriving subsystem and component properties, which are necessary for reaching defined full vehicle targets.

When analyzing the processes of the different disciplines, it becomes obvious that driving dynamics[2] already achieved a high progress in development of a structured and efficient process for cascading full vehicle targets to subsystem and component level by a wide application of simulative methods. However, in ride comfort the current process is less advanced (Rauh 2003, pp. 153–154), as virtual development predominantly relies on complex multi-body simulation models, which are not necessarily appropriate for early development phases. This is mainly attributed to the application and the necessity for parametrization of system properties, which are not required or available at the beginning of the property derivation process.[3] For the purpose of improving the process, a subsystem

---

[1]In this context, the automotive development process indicates the time frame in which a platform or vehicle project is completely developed, beginning at the definition of the product and ending at the Start-of-Production (short: SOP).

[2]Throughout this paper driving dynamics mainly refers to lateral dynamics respectively to the cornering behavior of the vehicle.

[3]For example, this can be the necessity of defining bushing stiffnesses to simulate with an multi-body component model, while the axle concept is still unknown in the early phase of the process.

methodology can be applied. However currently, subsystem parameters in ride comfort are not as clearly defined as in driving dynamics, so that existing abstract full vehicle models are based on them only to a limited degree. This is also a precondition for determining the dependencies of the full vehicle behavior from subsystem parameters. Therefore the scope of the following research mainly lies on integration of a respective level in ride comfort.

For that purpose, in Sect. 1.2 the state of the art in the automotive development process is shown. After examining the generic process, its specific state of realization in driving dynamics and ride comfort is analyzed. The analysis results in a determination of advantages in driving dynamics and an identification of deficits in ride comfort, which can potentially be resolved by applying a subsystem methodology. In Sect. 1.3 a modelling approach for simulating ride comfort on subsystem level is depicted. After describing general aspects, in Sect. 1.3.1 the most significant conditions for concept parameters on this level are derived based on the findings of Sect. 1.2. Afterwards specific parameters on subsystem level in ride comfort are presented. The integration of the presented modelling approach in the target cascading of the product development process is shown in Sect. 1.4. Beginning with targets of full vehicle development and therefore the definition of objective targets from subjective evaluation in Sect. 1.4.1, in the following Sect. 1.4.2 until Sect. 1.4.4 the derivation process from full vehicle over subsystem to component is depicted. In Sect. 1.4.5 the effects of the modified method on the development process are concluded. In the last section a summary of the research and an outlook will be given.

The objective goals of the current research are summarized as follows:

- Analysis of the Product Development Process with focus on driving dynamics and ride comfort concerning the derivation process
- Illustration of the structure of subsystem models in ride comfort
- Introduction of conditions for concept parameters on subsystem level and description of specific characteristics in ride comfort
- Demonstration, how a subsystem level can be integrated in the derivation process and description of the design process in general and with examples
- In this context, description of a method for determining objective targets of full vehicle development

1.2 Product Development Process

The product development process (PDP) of vehicles is characterized by high complexity and is based on deriving properties on different levels of detail of the vehicle. Mainly the process is represented by a V-model as described in ISO 26262 distinguishing between full vehicle, subsystem, and component level (Heißing et al. 2011, p. 496). A representation of the model is illustrated in Fig. 1.1.

Generally the process can be divided into two regions: target cascading, in which the concept development is conducted (left branch), and verification, in which the series development is carried out (right branch). In the first region, properties are

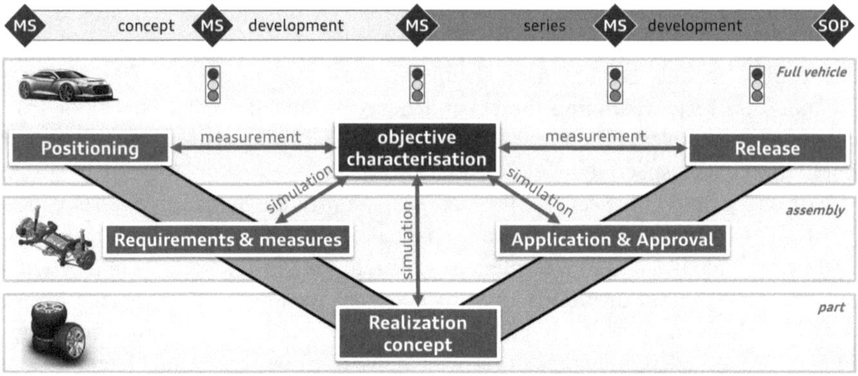

Fig. 1.1 V-model of the product development process of vehicles, adapted from Einsle and Fritzsche (2013, p. 750)

derived from full vehicle over subsystem to component level by providing development targets from lower to higher levels of detail. The assessed time period differs depending on the specifications of the vehicle manufacturer, but is usually located between product planning and concept freeze with a length of about 30 months. Concept freeze commonly takes place about 30 months before the Start of Production (SOP). However, the phases for derivation from full vehicle to subsystem as well as subsystem to component usually take about 3–4 months, meaning a short time frame for application of derivation methods.

In the verification area the developed components are assembled in simulation, but also tests on real vehicles are carried out by the series development. The targets defined in the cascading process are validated against the current values determined in the verification process, when analyzing the composition of components on subsystem and full vehicle level.

The described process is necessarily defined for different subsystems in full vehicle development, for instance suspension, tire, driveline or body but also different disciplines like driving dynamics, ride comfort, acoustics or durability (Heißing et al. 2011, p. 16). To meet new upcoming conditions like a higher model variety, higher functional demands, and reduced development times (Rauh 2003, p. 135) as well as new strategies like platform sharing, standardized modules, and shared parts (Heißing et al. 2011, p. 533), an efficient process needs to be continuously structured in and between these disciplines. Still the definition and sequence of procedures in the literature is relatively vague depending on the examined discipline.

At the beginning of the PDP in the target cascading process, a relatively high amount of unknown parameters exists in the early phase (Braess and Seiffert 2011, p. 899). However, the availability of simulation models in this period is desired so that frontloading (Hab and Wagner 2013, pp. 66–67 and 182–183) is enabled. Therefore throughout the process the share of applied simulative methods with respect to real tests is continuously rising to overcome emerging challenges of the automotive industry (Seiffert and Rainer 2008, pp. 7 and 73). In this case the effectiveness of the whole process depends on application and quality of simulation models (Bock et al. 2008, p. 11) by ensuring high functionality and reliability (Braess and Seiffert 2011, p. 902).

In the following, a short review of the state of the technology for driving dynamics and ride comfort concerning the PDP is given.

1.2.1 Driving Dynamics

In driving dynamics a high progress is already achieved in defining a structured development process with cascading and verification of vehicle characteristics. In this context the definition of objective vehicle characteristics has already been carried out (for example Decker 2009; Schimmel 2010) affecting the PDP in all phases. The obtained characteristics correspond to the targets of full vehicle development in the process depicted in Sect. 1.2 and establish the base for objective cascading of subsystem and component characteristics. In this context Schimmel has given a summary of determined objective criteria by using a steering wheel actuation model (Schimmel 2010, pp. 102–105) and refers to correlations between subjective evaluation and maneuver characteristics (Schimmel 2010, pp. 91–101).

The targets on full vehicle level are transferred on subsystem level using parametric concept models (Braess and Seiffert 2011, pp. 902–903). In driving dynamics typically single- and dual-track-models (Heißing et al. 2011, p. 95–105; Schimmel 2010, p. 25) are used for determining the contribution of different subsystems and their parameters on specific characteristics. A conventional dual-track model is depicted in Fig. 1.2.

Basically, in this modelling approach parameters on subsystem level are expressed by characteristic curves, like changes in wheel position due to applied forces, or characteristic values, like the location of the center of gravity or body mass. Therefore, conventional parameters for describing driving dynamics, like cornering stiffness or relaxation length, are implicitly or explicitly integrated. In particular the described approach has advantages when being applied in the development process, especially within the target cascading phase:

Independence of Concept
Considering axle and tire as black boxes, which are defined by parameters combining various effects, allows for a simulation without component properties in early phases of the process.

Simulation Speed
Due to the reduced set of parameters, computation times are decreased, enabling fast estimation of effects due to changes in parameters.

Analysis of Physical Relations
The lower complexity of the model results in a better overview over effects occurring due to interactions between different subsystems.

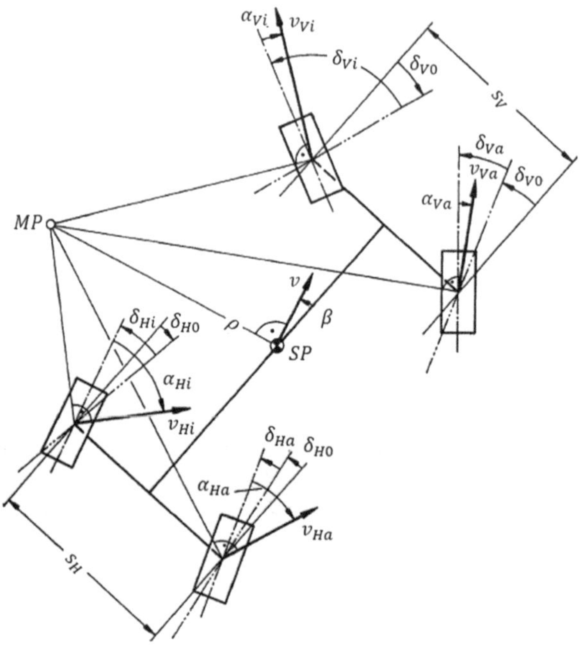

Fig. 1.2 Dual-track-model, adapted from Mitschke and Wallentowitz (2014, p. 834)

Fast Parametrization
Instead of measuring several components, the values of the simplified parameter space on subsystem level can be identified by measurements of the subsystem or full vehicle, which are for instance conducted on a kinematic and compliance test rig (Holdmann et al. 1998), which are less time-consuming.

Lower Error in Parametrization Process
The error of the addressed parametrization process is usually lower compared to the sum of errors of the component measurements, resulting in a higher quality of the simulation.

Option for Parametrization of Competitor's Vehicles
Due to the faster parametrization process compared to the process on component level, a parametrization of any car is enabled in a limited time frame, allowing for an analysis of competitor's vehicles.

The mentioned advantages are now able to contribute to a structured process in driving dynamics, resulting in benefits in defining objective targets on full vehicle level and deriving properties on subsystem and component level.

1.2.2 Ride Comfort

In contrary to driving dynamics the state of technology in the development process of ride comfort is less advanced (Rauh 2003, pp. 153–154). This is due to the fact, that the system dynamics and the identification of comprehensive ride comfort targets is far more complex than in driving dynamics as well as the subsystem level is not clearly defined. The ride comfort targets will be addressed in Sect. 1.4.1, being the base for derivation of vehicle characteristics. The definition of the subsystem level will be in focus throughout the remaining research.

Basically, the derivation process in ride comfort is performed using models on component basis, typically integrated in a multi-body simulation (Heißing et al. 2011, pp. 504–506). Therefore, a transfer of development targets from full vehicle level directly to component level becomes necessary. However, specific conditions concerning models on component level, which are associated with the development process depicted in Sect. 1.2, influence this procedure:

Availability of Predecessor
Dependent on the intended vehicle, a predecessor not necessarily exists, meaning that no concept can be used as basis. For instance an axle concept cannot be presupposed in this case. If a predecessor exists, still another concept can be more suitable for the successor.

State of Predecessor
If a predecessor exists, the multi-body model may be not up-to-date in practical applications, since not all changes in late phases of the development process are integrated in the simulation model. Therefore, an additional time frame for updating the model has to be provided.

Availability of Component Parameters
In early phases of the development process component parameters are mostly unknown, preventing the simulation of ride comfort characteristics.

Design by Trial-and-Error
Due to the high complexity of component models, also a high amount of parameters can be set, which additionally have interdependencies when influencing subsystem or full vehicle properties. Therefore, the design is predominantly based on trial-and-error (Heißing et al. 2011, p. 502). A structured derivation of measures for reaching full vehicle targets is impeded.

Reliability of Defined Data
The reliability of available parameters is differing. Since some parameters like masses can change more frequently throughout the process, a fast estimation of consequences on full vehicle level is necessary.

Limited Time for Defining Actions
Related to the upper element, designer on component level need procedures to influence vehicle behavior in specific ways in a limited time period when designing component models.

Considering mentioned conditions the application of component models, especially of higher complexity with long simulation times, becomes more difficult. Compared to the advantages of models on subsystem level in driving dynamics (cf. Sect. 1.2.1) direct correlations for resolving occurring issues become evident. Therefore further proceedings of this research focus on models enabling the same advantages of subsystem models in ride comfort.

For this purpose, in the next section a modelling approach is introduced, which is specifically developed for the integration on subsystem level in the product development process, allowing for an efficient derivation process.

1.3 Models for Simulating Ride Comfort on Subsystem Level

In the following section, a modelling approach for ride comfort on subsystem level will be depicted. Afterwards in Sect. 1.4 the application of the model will be integrated in the target cascading process of the V-model between full vehicle and component.

Concerning the description of the modeling approach, it has to be noted that for a derivation of the detailed model structure, further extensive analyses have to be conducted, whose illustration are beyond the scope of this paper. However, the model structure is outlined by its basic principles.

The precondition for the modelling approach is to reduce the component properties of different subsystem concepts (like suspension or powertrain mountings) on common properties on subsystem level. At the maximum degree of abstraction these become black boxes, having specific inputs and outputs, like deflections or forces. Their interior shall only be known during development of the model, but is neglected during application in the process for maintaining independence of concept. The resulting individual subsystem models are connecting bodies with aggregated mass and inertia properties among each other.

A representation of the concept implemented in a full vehicle approach is given by Fig. 1.3.

In the modelling approach, the excitation by the road profile is given at the wheel contact patch, which is defined separately for the four wheels and is transferred by a tire model to the corresponding tire-sprung masses. From this location the information is transferred to the remaining bodies like vehicle body, subframe or engine by similar subsystem models reproducing the properties of the respective mountings. Instead of using component parameters, the properties of the components are summarized in general stiffnesses, like a longitudinal stiffness of the subframe or

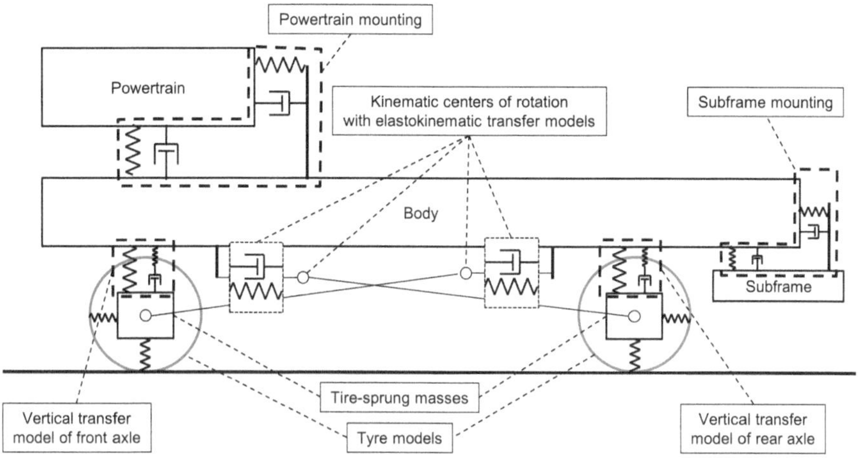

Fig. 1.3 Approach for a full vehicle concept model on subsystem level

powertrain mounting. Generally, the degree of freedom of the individual subsystem masses is six, but has to be reduced for considering only most relevant parameters. A more detailed description of the subsystem structures is given in Sect. 1.3.2.

Based on this approach, conditions for the selection of concept parameters defining the transfer behavior of these elements are presented in the following. Subsequently, the requirements are applied on ride comfort models by introducing specific parameters on subsystem level.

1.3.1 *Conditions for Concept Parameters on Subsystem Level*

The advantages of models on subsystem level in driving dynamics, depicted in Sect. 1.2.1, combined with the boundary conditions given by the development process, presented in Sect. 1.2.2, serve as a basis for defining requirements for concept parameters on subsystem level in ride comfort. Beyond that, conditions enabling a structured integration in the development process are given.

Dominant Influence on Ride Comfort Targets

As concept models on subsystem level aim on being as simple as possible while still maintaining a sufficient quality of a prognosis, properties having a significant influence on ride comfort targets have to be integrated while remaining parameters are neglected. Thereby, a fast application, parametrization, and flexibility of the model is maintained.

For example, in the frequency range of body vibration phenomena, the properties of the damper mainly define the dynamic behavior of the suspension in vertical direction while contribution of elastomer bushings to the damping rate can be neglected.

Relation to Subsystem Level
Defined parameters need to be specified on subsystem level as given by the previously described approach for maintaining the independence of a concept. Therefore, in early phases of the development process, simulation without knowledge of the subsystem concept or component parameters becomes possible.

In this context, the integration of characteristics such as the overall longitudinal stiffness of a suspension is convenient while for instance the stiffness, position, and orientation of a single rubber mount is inappropriate.

Availability and Reliability of Parameters in the Development Process
Due to specification of vehicle properties at different times in the PDP, it becomes necessary that used parameters are available at the beginning of the concept phase at subsystem level or are easy to identify through test rig measurements in a short time frame, as depicted in Sect. 1.2. Only if these parameters are known, the reliable application of the model in short time frames of the early process phases is enabled.

For instance, the distribution of the vertical wheel load is determined in early phases of the process, while on the other hand, the specific masses of components (like wheel carrier, spring strut or transmission) are still not available.

Relation to Parameters Typically Used in the Process
For practical purposes in application of the model, predominantly characteristic parameters established in the development process shall be used. Thereby an efficient process due to improved handling and communication, when using the model, is ensured.

In this context, the positioning of instantaneous centers of rotation as well as the specification of support angles[4] is reasonable, while cross-terms[5] in the suspension transfer matrix are currently not well established in the development process of suspensions.

Correlation to Other Models
In application of various models in other disciplines, an efficient process is ensured when parameters are similar, allowing for a likewise application of different models in the same task or the combination of modelling approaches.

For instance, if one model defines support angles while the other uses instantaneous centers of rotation, the comparability between these methods, while ensuring the application of the same parameters is impeded.

Considering these requirements for concept parameters on subsystem level, in the next section specific parameters in ride comfort models are described.

[4]Support angle means the angle defining the amount of vertical force which occurs due to longitudinal or lateral forces on a suspension, predominantly defined by its kinematics.

[5]In this context cross-term means parameters not lying on the main diagonal of the transfer matrix and defining the reaction of the system in another degree of freedom than in the direction of the excitation, which correlates with support angles.

1.3.2 Concept Parameters on Subsystem Level in Ride Comfort

In the following, the application of the referred parameters is carried out under the conditions mentioned in the previous section. As already mentioned in Sect. 1.3, a derivation of the detailed structure of the subsystems is beyond the scope of this paper, but a summary, illustrating the basic principles, is given for the individual systems.

When examining ride comfort characteristics in the given research, the frequency range from zero till 30 Hz is observed. Therefore the vibration of vehicle body, engine, tire-sprung masses and subframe as rigid bodies are of particular interest. Furthermore, natural frequencies of the body structure can occur, but shall be neglected in the investigation. Based on these conditions, in the following the subsystem behavior of tire, suspension, and the mountings of subframe and powertrain need to be modelled. In the current research, the analysis is predominantly performed with focus on the suspension. For the remaining subsystems, conditions for developing an appropriate subsystem approach are given.

The tire, being subsystem and component at the same time, is usually represented by a single-point contact model for long wavelengths occurring for instance at natural frequencies of the body. At higher frequencies shorter obstacles are enclosed (cf. Fig. 1.4) requiring a more complex modelling approach. Therefore at low frequencies the predominant tire property on subsystem level is the overall vertical tire stiffness while with rising frequency respectively shorter wavelengths, longitudinal stiffness and geometrical aspects of the tire are getting more important.

With respect to the defined frequency range, a tire model needs to be used, which allows to reproduce the enveloping properties of the tire and which can be parametrized on a tire test rig in a short time frame, like MF-SWIFT (Pacejka 2006, pp. 412–510).

The vehicle suspension serves as interface between tyre-sprung mass and body respectively the subframe, if latter is mounted elastically on the body. The transfer behavior of the subsystem can then be defined by static and dynamic stiffness in all six directions in space, forming a 6-by-6 matrix with variable coefficients for reproducing dynamic properties. However, as already described in Sect. 1.3.1, instead of using cross-terms of the matrix, a more sophisticated method of abstraction is applied by dividing the transfer behavior into a diagonal stiffness matrix and separate kinematic properties. In this context the stiffness matrix incorporates

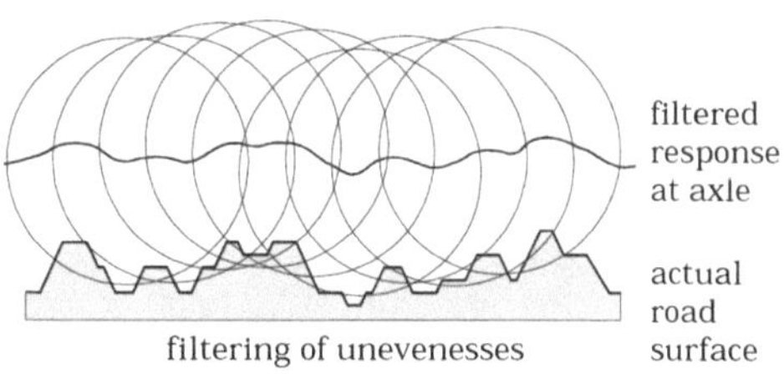

Fig. 1.4 Filtering of unevenness of a tire as depicted in Zegelaar (1998, p. 58)

elasto-kinematic properties (for example in longitudinal and lateral direction) as also a stiffness in vertical direction, which is usually attributed to kinematics. On the other hand, the separate kinematics avoids the application of cross-terms in the stiffness matrix by using geometrical relations.

This is done for every connection of tire-sprung mass and body, but also for alternate movement of tire-sprung masses between left and right wheel, if necessary for the respective direction. Additionally every element of the stiffness matrix is wheel-based, meaning the relation is defined between force and displacement at the same location on the wheel, which maintains the independence of axle geometry respectively lever ratios.

Under the described conditions for dividing the subsystem model of the suspension into a diagonal stiffness matrix and kinematics, different concept parameters can be identified.

The overall vertical stiffness of the suspension affects body accelerations over a wide frequency range, being involved in quite all maneuvers relevant for ride comfort, beginning at the natural frequencies of the body. The parameter combines the stiffness of main spring, torsional stiffness of bushings[6], and the bump stop (Bindauf et al. 2014, p. 78).

Therefore also the vertical damping of the suspension can be defined as important parameter on subsystem level, affecting the reaction of the axle due to dynamic excitation. In this case, the components contributing to the summarized damping force can be identified as the same as for vertical stiffness. Still, it can be assumed, that the influence of the damper dominates the force generation over a wide frequency range, so that in most cases damping due to torsional deformation of bushings[7] can be neglected. With rising frequency also the damper top mount has to be considered (Bindauf et al. 2014, p. 80).

When the wheels respectively the tire-sprung masses of an axle are unequally deflected in vertical direction, an alternate vertical stiffness comes into effect. The influence can be modelled by defining a wheel-based stiffness as coefficient of vertical force and differential deflection acting between the tire-sprung masses and the respective body connection. Predominantly, the properties of the anti-roll bar are responsible for this effect.

In a similar manner an overall stiffness in lateral direction of the axle can be defined. Therein predominantly the stiffness of bushings is included. A parameter combining several individual damping properties of the bushings can be defined as well. While in driving dynamics a high lateral stiffness is important for maintaining the wheel position when lateral forces are applied (Heißing et al. 2011, p. 456), the influence of both mentioned parameters on ride comfort is mostly unknown.

In longitudinal direction also an overall stiffness and damping can be defined. Thereby the individual locations, orientations and properties of the involved

[6]The wheel-based stiffness due to bushings is usually called secondary spring rate, probably being mainly dependent on the torsional stiffness of bushings.

[7]Analogue to the secondary spring rate this effect will be called secondary damping rate.

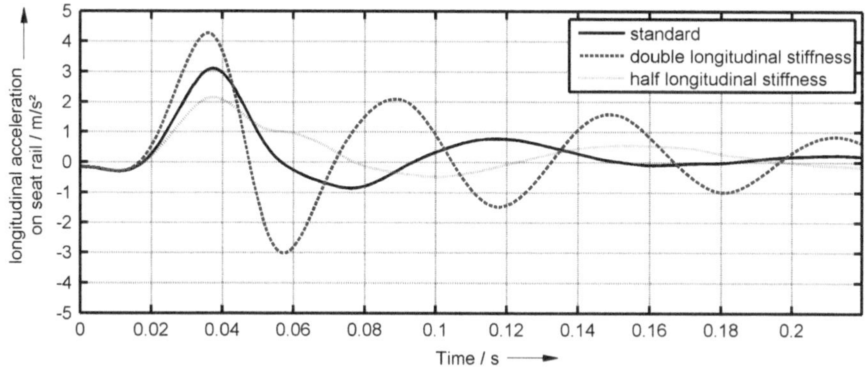

Fig. 1.5 Influence of longitudinal stiffness on seat rail acceleration in longitudinal direction when passing a cleat

bushings are abstracted and an approach independent of the suspension concept is generated. The overall longitudinal elasticity of the suspension comes into effect when the tire generates longitudinal forces as a result of the road profile. It can be assumed, that this predominantly occurs with rising frequency of the excitation. Therefore, the influence on ride comfort can directly be deduced when analyzing associated maneuvers with the help of an appropriate concept model. For example, in Fig. 1.5 the influence of longitudinal stiffness on seat rail acceleration, when a cleat is passed, is depicted.

As can be seen in the figure, a higher longitudinal stiffness results in a higher longitudinal peak acceleration, a higher vibration frequency and a longer decay process. This occurs due to a higher resistance of the axle in longitudinal direction, when the tire is passing the obstacle and an associated decreased effectivity of the damping.

In longitudinal and lateral direction also an alternate stiffness can be defined. Though, the relevance of these effects depends on the usage of coupling elements in the suspension, like a subframe. For instance, without subframe the alternate longitudinal stiffness of an axle can be neglected during static maneuvers, when certain conditions are met (Bindauf et al. 2014, p. 79).

Beside the response of a suspension due to longitudinal and lateral excitation in the same direction, a coupling between these directions and the vertical direction is generated by axle kinematics. At subsystem level, this behavior can be described by a support angle or an instantaneous center of rotation (Matschinsky 2007, pp. 23 and 48). Considering their dependence of vertical, longitudinal and lateral wheel deflection, they can serve as subsystem parameters. On the one hand, these kinematic properties characterize the amount of vertical force generated by longitudinal and lateral forces on the wheel, on the other hand they define the kinematic movement of the tire-sprung mass with respect to the body (Matschinsky 2007, p. 41). Therefore an additional vertical force occurs when longitudinal forces are generated in specific maneuvers, as described before when addressing longitudinal stiffness. The behavior in lateral direction can be described similar, but corresponding maneuvers are different. As currently cornering is mostly not considered in maneuvers defining ride comfort, lateral forces predominantly are generated during compression and rebound of the suspension, as a consequence of changes in track and toe or when obstacles are asymmetrically enveloped by the tire with respect to the wheel center plane.

Another effect which influences suspension response, but can generally not be attributed to stiffness or damping properties is static axle hysteresis or axle friction. The effect has been researched to some extent in literature (Yabuta et al. 1981; Gillespie 1992, pp. 166–168; Nakahara et al. 2001), but first investigations concerning finding an integral approach for modelling, designing and parametrization of axle friction in the development process are given by Angrick et al. (2015, pp. 377–403).

When the suspension includes an elastically mounted subframe, additionally it becomes necessary to consider the connection between subframe and body. For this purpose, the stiffnesses and dampings of the individual components are summarized in generalized stiffnesses and dampings for the whole mounting. As the properties in longitudinal and vertical direction are of main importance for ride comfort, focus lies on the associated parameters.

The powertrain mounting is abstracted with the same method, but as the coupling of the different degrees of freedom is generally more complex (due to asymmetric stiffness and hydraulic properties of the bushings) than in the subframe mounting, the described method is only partially applied by disregarding less important parameters in the stiffness matrix and maintaining the hydraulic properties on component level.

Individual bodies, which are connected by mentioned different mountings, are represented on subsystem level by summing up mass and inertia properties of related components. In this case, properties, which are irrelevant for ride comfort, can be neglected, like yaw inertial torque of the vehicle body or of tire-sprung masses. As a consequence, the aggregated mass and inertia properties of the body correspond to subsystem parameters. This enables the possibility for parametrizing whole subsystems in the development process, when component parameters are still unknown. In particular, this concerns tire-sprung masses (separately for each side of front and rear axle), powertrain, vehicle body as well as subframe and differential, if necessary.

1.4 Integration of a Subsystem Level in the Derivation Process from Full Vehicle to Components

After introducing a modelling approach on subsystem level, in the following section, the target cascading process of vehicle characteristics in ride comfort from full vehicle to component level will be presented, integrating a subsystem level in the PDP. The process will predominantly be exemplified on the suspension.

First, in Sect. 1.4.1 targets of full vehicle development will be described concerning the derivation of objective targets from subjective evaluation. Based on the results, the derivation from full vehicle to subsystem level is depicted in Sect. 1.4.2. The significance of the subsystem level for the development process is presented in Sect. 1.4.3. Using this as a basis, the cascading to component level is described in Sect. 1.4.4. In the last section the findings are summarized.

1.4.1 Targets of Full Vehicle Development

The subjective evaluation of ride comfort is a key method in the development process of a car. However, this method cannot be used in early stages of development as prototypes are still not available. Therefore, objective and computable criteria for ride comfort are needed. Such objective criteria, often in the form of characteristic values, should be related to relevant subjective criteria. A comprehensive collection of subjective evaluation criteria is shown in Fig. 1.6.

Representative characteristic values for ride comfort should fulfil different requirements:

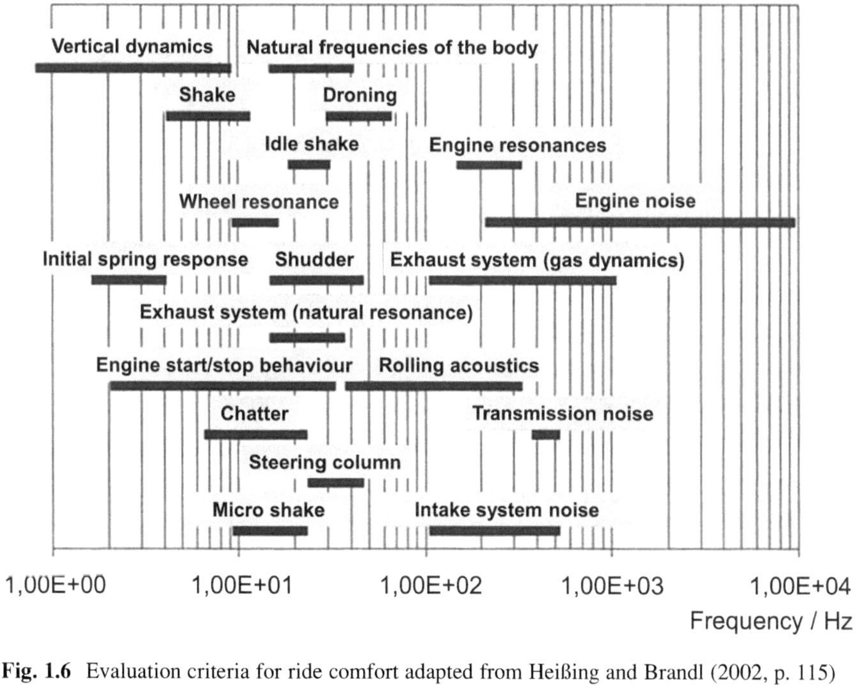

Fig. 1.6 Evaluation criteria for ride comfort adapted from Heißing and Brandl (2002, p. 115)

Computability
The characteristics shall be computable out of a simulation.

Measurability
Under real testing conditions, the values need to be easily identifiable out of measured signals from the full vehicle.

Completeness
The relevant ride comfort criteria have to be described by the targets completely and comprehensively.

Assignability
The transfer from subjective to objective criteria needs to follow a distinct process.

A literature research results in a huge amount of possible descriptions for ride comfort. Many of these approaches are based on correlation of subjective and objective evaluation, either with analytical weighting methods (Cucuz 1993; Hennecke 1994; Klingner 1996; ISO 2631) or with representation of the unknown correlation by neuronal Networks (Albrecht and Albers 2004; Stammen and Meywerk 2007).

The usage of these values is often restricted to selected contact points and corresponding directions between driver and vehicle or to specific excitation profiles. Additionally many approaches summarize ride comfort in one value, which does not correspond to conventional testing methods in subjective evaluation and therefore not meets the mentioned requirement of assignability.

Therefore many OEM[8] do not work with such approaches and use specific, not weighted values for the description of relevant and standardized excitations instead.

An example could be the description of the response of a car while passing a cleat, as shown in Fig. 1.7. In this case, the hardness and the decay behavior due to the impact are of particular interest in subjective evaluation. Characteristic values can be defined by the Peak-to-Peak value (*P2P*), the vibration frequency (f_d) as well as the decay constant (δ) resulting out of the measured acceleration at the seat rail of the driver. These values fulfil the requirements above and can be monitored along the development process of the car.

Apart from the behavior of the car when passing a cleat, further characteristic values can be defined. These differ with respect to the operational methods of each OEM, but still the classification is usually carried out in similar categories. In detail this concerns the frequency dependent transfer behavior of the body, the previously described step response or the response on stochastic roads. Similar to Fig. 1.7, characteristic values can also be defined in these cases. For instance, the body response over frequency can be separated in the range of its natural frequencies and in corresponding resonances of subsystems at higher frequencies. For stochastic roads effective values, like root mean square values (RMS), can be defined. In the product development process, these characteristic values are then used as a basis for a derivation of properties.

[8]OEM: Original equipment manufacturer; corresponding to the common definition in the automobile sector, the term OEM means manufacturers of vehicles, selling them under their own brand.

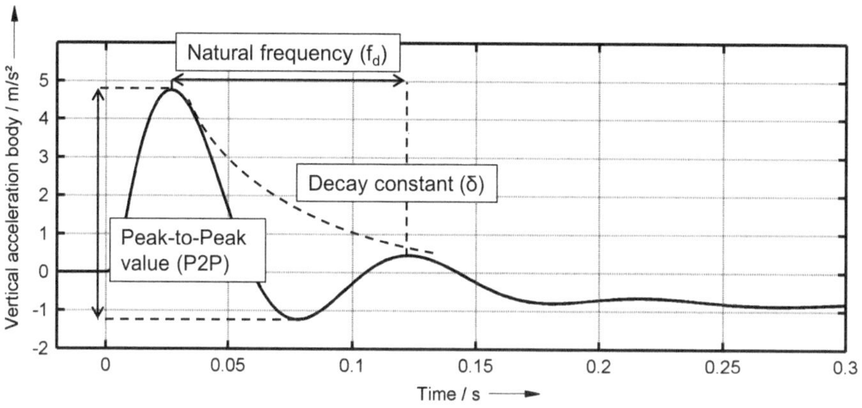

Fig. 1.7 Step response of the body when passing a cleat

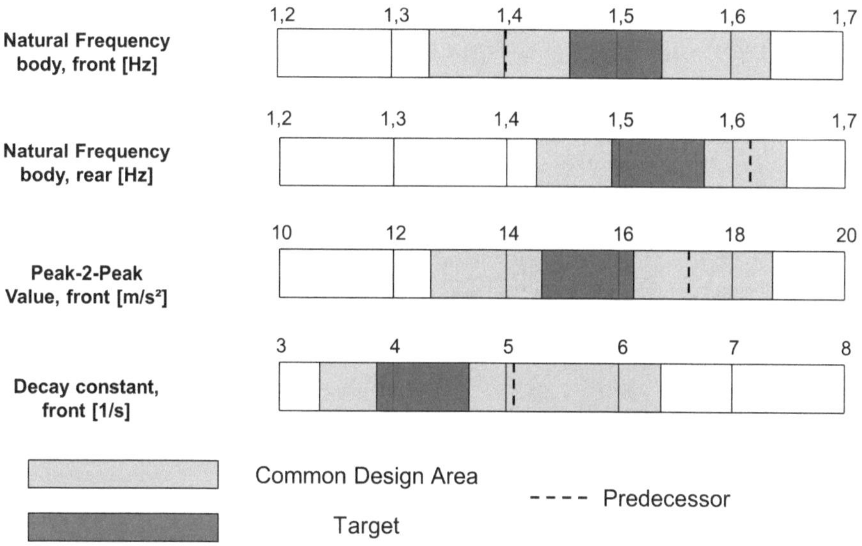

Fig. 1.8 Definition of targets of full vehicle development for different characteristics

For this purpose, primarily a comparison of current vehicles in one class of different brands is conducted and characteristic values are determined. Depending on the differences between competitor's vehicles and the OEM's own vehicles, advantages and disadvantages are identified. Subsequently, new target values for a successor or a corresponding new vehicle class are defined. The relationship between predecessor and new target values can be depicted using a bar diagram, shown exemplary in Fig. 1.8.

At this precise moment, models on subsystem level already can support the development process as competitor's vehicles can also be measured on subsystem test rigs, for instance on a kinematics and compliance test rig, while the determination of component properties is not practicable in a limited time frame. The parametrized models allow for analyzing the contribution of determined parameters on subsystem level to the behavior on full vehicle level, supporting the basis for decisions concerning the derivation process.

In the present figure, natural frequencies of the body on front and rear axle as well as Peak-to-Peak and decay values have been defined as target values. Concrete values of the predecessor are visualized by a dashed line, while target values for the new vehicle are represented by a range of possible values, allowing for a tolerance when designing parameters. Additionally common areas within these values are depicted, which depend on the conditions of the particular criterion and vehicle class.

Overall the depicted process provides a possible approach for a transparent and integrated development process based on full vehicle characteristics which can be transferred to different objective criteria of subjective evaluation for ride comfort.

1.4.2 Deriving Properties from Full Vehicle to Subsystem Level

In the following section, the full vehicle targets described in the previous section are derived using the model approach given in Sect. 1.3.

The basic principle for the cascading process is to define full vehicle targets as reference when varying subsystem characteristics. In this context, a useful approach is to design subsystem parameters from lower to higher frequency with respect to the influence of characteristics to a corresponding phenomenon. Thereby procedures can be found for defining individual parameters within the scope of different full vehicle targets. However, boundary conditions concerning interdependencies within the same discipline and other disciplines need to be considered. By parallel application of models simulating characteristics of all disciplines, an optimum on full vehicle level can be found.

The procedure will be exemplified on vertical and longitudinal stiffness of the suspension. As already described in Sect. 1.3.2 the vertical stiffness influences ride comfort in a wide frequency range. Starting at low frequencies this shall be one of the first parameters to be designed. The representation in a subsystem model can be carried out with a characteristic curve, as shown in Fig. 1.9.

The curve can be divided into three main regions: rebound stop, linear region and progressive region. In this case, the region of the rebound stop is considered as irrelevant for ride comfort, as the deflection amplitudes in associated maneuvers are too small for reaching this area. The suspension predominantly operates within the subsequent linear section. In this case, linear corresponds to a constant increase of the vertical force related to the vertical deflection. With higher amplitudes additionally a progressive increase becomes obvious in the curve, limiting the maximum deflection when compressing the axle. Under these conditions, two boundaries for specifying the

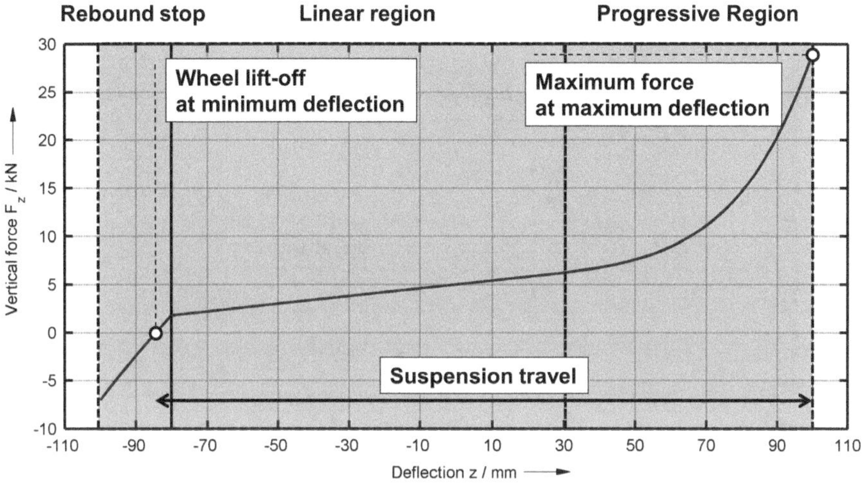

Fig. 1.9 Significant properties describing the shape of the characteristic curve between vertical force and vertical deflection

design area of the suspension in vertical direction can be found. At zero force the wheel lifts off the ground, limiting the vertical deflection in rebound direction. In direction of compression a point can be defined where a specific maximum force occurs at a maximum deflection, predominantly specifying the desired limit for axle compression. Both values define the suspension travel in vertical direction, being an important criterion on subsystem level for defining vertical stiffness. Within this range an optimum curve for reaching full vehicle targets needs to be defined.

As wheel load distribution is provided in early phases of the development process, in the linear region a constant stiffness can be found by changing the parameter until the natural frequency of the body defined on full vehicle level is reached. In this context the damping can also be varied for reaching the defined amplitude of the vibration. The procedure is conducted for front and rear axle.

However, interdependencies between full vehicle targets in the same discipline as also other disciplines need to be considered, when conducting this procedure.

When changing a parameter having differing effects on characteristic values, trade-offs are occurring. Within ride comfort the magnitude of the transfer function between body and road when varying damping characteristics can be used as an example, as shown in Fig. 1.10.

With rising damping the magnitude of the transfer function decreases in the resonance area and increases in the isolation area. This trade-off needs to be optimized based on full vehicle properties for an optimum solution. As the deflection amplitude of the suspension generally changes with the excitation frequency under operating conditions, an important factor for resolving the shown trade-off can be found in the nonlinearity of the damper curve, which allows the definition of different damping ratios at various amplitudes.

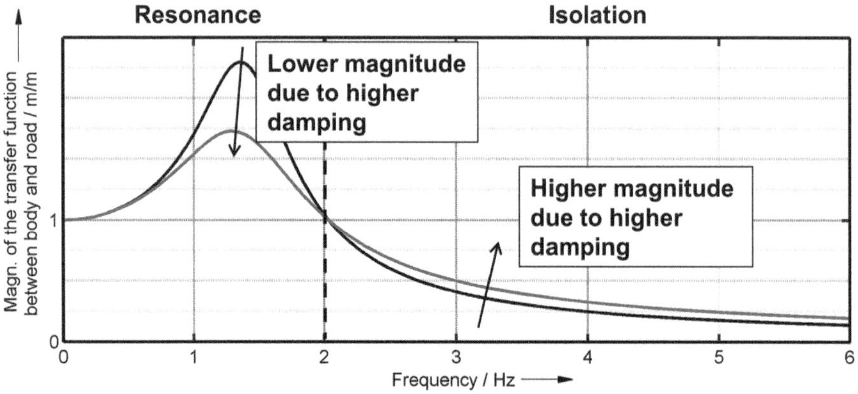

Fig. 1.10 Influence of damping on the magnitude of the transfer function between body and road

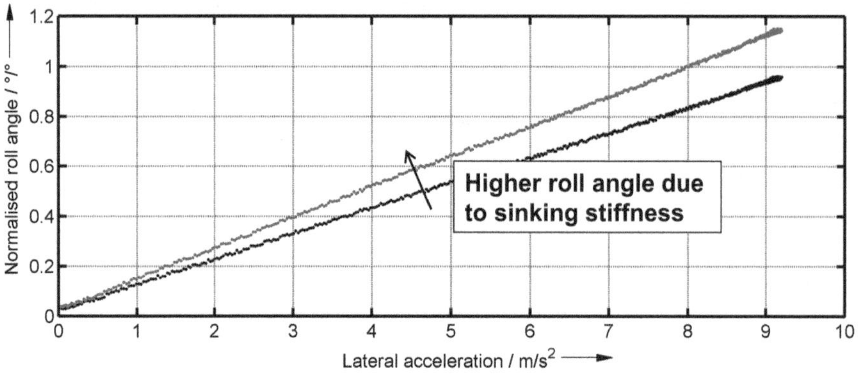

Fig. 1.11 Influence of vertical stiffness on roll angle gradient in driving dynamics

On top of that, the influence on full vehicle characteristics of other disciplines needs to be considered. In this case, the correlation to other models described in Sect. 1.3.1 is of importance as the parametrization should be similar for ensuring maximum comparability of the results.

An example is given by the influence of vertical stiffness on ride comfort and driving dynamics. Decreasing stiffness reduces coupling between wheel and road which partially results in better ride comfort. In driving dynamics on the other hand the roll angle gradient is rising with lower vertical stiffness, as shown in Fig. 1.11. This correlates with a lower score in subjective evaluation of driving dynamics for this characteristic value. Additionally changing load has a higher impact on ride height and the operating point is shifting towards the progressive region, depicted in Fig. 1.9. By modifying the alternate vertical stiffness, the effects in driving dynamics can partially be resolved. However, this also causes higher dynamic roll accelerations when driving over asymmetric road profiles.

Under the described conditions the vertical stiffness can be designed independently of longitudinal stiffness, whose influence on the vertical body natural frequency can be neglected. As already described in Sect. 1.3.2, the longitudinal stiffness comes into effect when the tire transfers longitudinal forces into the suspension, which is a phenomena at higher frequencies in ride comfort, for instance when a cleat is passed. In this case, longitudinal stiffness influences the step response in longitudinal direction while the support angle between vertical and longitudinal direction transfers an additional force in vertical direction. As the vertical stiffness of the axle has an important influence on the step response in vertical direction, the parameter also needs to be designed for this application. Therefore after designing vertical stiffness for low frequency phenomena, as described above, the parameter needs to be optimized with respect to phenomena at higher frequencies as well, to find an optimum solution.

In addition to vertical and longitudinal stiffness, the described procedure is conducted as already proposed with rising frequency incorporating an increasing amount of parameters and full vehicle targets.

1.4.3 Subsystem Level

By applying the method depicted in Sect. 1.4.2, the contribution of different subsystems for reaching full vehicle targets can be analyzed. This is shown on the example of the natural frequency of the body at the front axle, depicted in Fig. 1.12.

In this example the target natural frequency for the vehicle is increased for realizing a more distinct differentiation between vehicle classes and having benefits in driving dynamics, while accepting partially worse ride comfort. A lower body mass has been specified by means of fuel consumption, but still contributes for reaching the target. Also the vertical damping has been specified separately, as it has more influence on other targets, but partially shifts the natural frequency. By

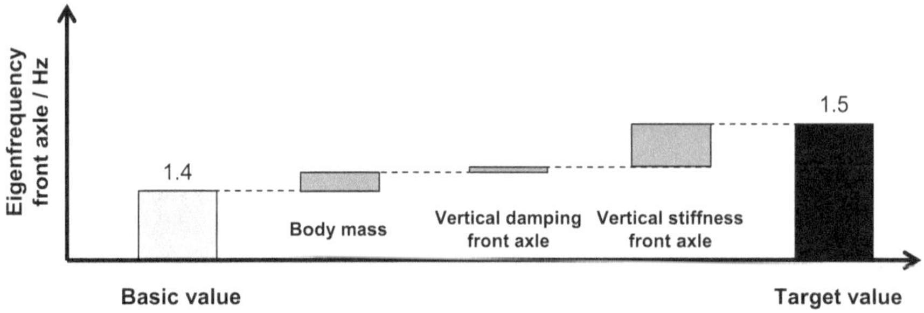

Fig. 1.12 Contribution of different subsystems for reaching the natural frequency on the front axle of the body as defined on full vehicle level

rising the vertical stiffness, as dominant property for influencing this characteristic, the remaining gap is eliminated.

Even when examining the suspension individually, it can be observed that several parameters within one subsystem, in this case vertical stiffness and damping, influence multiple full vehicle characteristics. Therefore the given process can be defined as multi-input multi-output system (Einsle and Fritzsche 2013, p. 758).

As a result of the analysis, it is possible to define the subsystem level as a new reference between full vehicle and component. From now on, the concept parameters on this level, described in Sect. 1.3.2, serve as targets for the component level. The benefits of this method will be addressed in Sect. 1.4.5.

1.4.4 Deriving Properties from Subsystem to Component Level

Using the subsystem properties as a new reference, generally a similar derivation process as described in Sect. 1.4.2 can be conducted between subsystem and component. At this moment a pre-selection of subsystem concepts can be carried out, which is not possible between conventional approaches, acting between full vehicle and component.

The pre-selection is performed by comparing derived properties on subsystem level with those which are characteristic for several subsystem concepts. For instance, Heißing et al. give an overview over different axle concepts and their advantages as well as disadvantages (Heißing et al. 2011, pp. 421–459). When such characteristic properties are expressed as possible ranges of parameter values on the basis of objective criteria, a comparison to determined subsystem properties for reaching full vehicle targets and therefore a pre-selection of a concept becomes possible. This will be illustrated on an example.

As already described in Sect. 1.4.2, suspension travel is an important parameter for defining the area, in which the vertical characteristic curve is designed. When analyzing the required suspension travel, determined by application of subsystem models, axle concepts can potentially be excluded. When assuming the suspension travel for reaching full vehicle targets needs to be maximized, a multi-link suspension is favored instead of a spring-strut-type axle, as the former has less demand in height of the construction and therefore allowing for a higher suspension travel at the same height. Additionally a higher flexibility in the design of kinematic parameters is given by the multi-link suspension, if necessary for reaching full vehicle targets. On the other hand the spring-strut-axle can be favored when a high longitudinal elasticity is needed.

These analyses need to be conducted for all subsystem parameters concerning full vehicle targets in every discipline (e.g. ride comfort or driving dynamics), also considering costs of the respective concept. By considering this, design of concepts on component level, which have been excluded on subsystem level, can be neglected, preventing the selection of an inappropriate variant and therefore resulting in a higher utilization of the potential for finding an optimum solution.

After pre-selecting a desired concept, the derivation of subsystem to component level is conducted. A multi-body simulation model is developed including components for representing the particular system. In case of the suspension this would be for example the stiffness, damping, mass, location, and orientation of spring, damper, levers, and bushings. The determination of component parameters is conducted by using optimization algorithms (Heißing et al. 2011, p. 502). An advantage is given by having a direct reference on subsystem level, allowing for comprehension of effects resulting out of changes in component parameters. By referencing on full vehicle level, processes are far more complex, particularly making the analysis more difficult or irresolvable.

The derivation process will be exemplified on the vertical characteristic curve of Sect. 1.4.2. As already described in Sect. 1.3.2, the vertical stiffness is mainly affected by the stiffness of spring, bushings, and bump stop combined with spring and damper ratio. The ratios result of the geometrical positioning of these components and the length of corresponding levers.

At this moment, multiple solutions can be found for representing the wheel-based overall vertical stiffness, initially resulting in an under-determined system. However, also on this level interdependencies to other disciplines have to be considered. While compromises between driving dynamics and ride comfort have been found on subsystem level to a great extent, displacement and cutting forces in components can now be determined. In this case, interdependencies to the geometrical package, durability, production, and assembling arise. For instance the properties of the progressive curve defined by the bump stop are verified concerning productability of the running-in characteristic as well as the durability of the component. If the selected configuration is not feasible, connecting points need to be shifted, resulting in a changed geometry and cutting forces. A second example would be stiffness and positioning of spring and levers, which need to enable enough clearance for preventing collisions of components while still maintaining the same wheel-based stiffness. Considering these conditions along with the remaining subsystem characteristics, it is assumable that the system becomes over-determined, allowing for application of optimization algorithms on different criteria.

After designing properties on component level, these are still integrated in a full vehicle multi-body model for verifying targets on all levels. Subsequently, the contribution of different component properties on subsystem and full vehicle level can be defined similar to the process shown on basis of Fig. 1.12.

1.4.5 Benefits in the Derivation Process Using a Subsystem Level

In the previous sections the derivation process of vehicle characteristics has been shown on the example of ride comfort. Finally, the dominant benefits when integrating a subsystem level in the PDP will be concluded:

Higher Utilization of Potential in Concept Selection
By determining parameters on an intermediate level between full vehicle and component, a comparison between required subsystem properties and characteristics of different concepts is enabled. Therefore the selection of an inappropriate concept can be prevented and the probability of finding an optimum solution is increased.

New Reference in Derivation
When designing component parameters, a direct relation on subsystem parameters becomes possible, allowing a reduction of complex interdependencies, typically occurring when relating on full vehicle level.

Decreased Complexity Concerning Trade-Offs
As trade-offs concerning driving dynamics and ride comfort can be partially resolved on subsystem level, the design of components considering subsystem characteristics contributes to a solution in both disciplines. Thereby the complexity when treating additional interdependencies on component level is reduced.

Individual Examination of Effects
The properties defined on subsystem level can be changed independently of each other. Therefore the impact of changes in properties on full vehicle targets can be examined individually for every parameter, which contributes to the understanding of the system. On the other hand, a change in properties on component level results in a variation of multiple subsystem properties, impeding for comprehension of effects on full vehicle level, if an examination on subsystem level is neglected.

1.5 Summary and Outlook

Within the present investigation, the design of ride comfort characteristics on subsystem level in the development process has been depicted. For that purpose, the product development process has been analyzed with focus on driving dynamics and ride comfort, resulting in specific differences concerning the process, which are applied in both disciplines. While driving dynamics already achieved a high progress in defining a structured and efficient procedure, the derivation of properties in ride comfort is less advanced. Deficits in the corresponding process are pointed out.

Due to this matter, the research focuses on the application of subsystem models in ride comfort. Therefore, the structure for developing adequate models in this discipline is given. The parameters for enabling an appropriate derivation process need to fulfil multiple requirements, which are summarized in Sect. 1.3.1. Specific parameters in ride comfort are depicted in the subsequent section.

Afterwards, the models are integrated in the derivation process in ride comfort, where initially a novel method for determining objective targets of full vehicle development is described. Requirements for representative characteristic values are depicted and a definition of specific, not weighted values is given generally and on the example of a cleat excitation. Differences to previous methods are described.

The resulting criteria serve as a basis for deriving properties on subsystem and component level. In the derivation process, the applied method is based on the design of parameters from low to high frequency ranges. By considering the differing influence of the given subsystem parameters on full vehicle characteristics and the interdependencies in ride comfort as well as other disciplines, a structured derivation on subsystem level is enabled. In the following, component parameters are derived and differences to the process singularly relying on full vehicle targets are presented. The analysis shows that the integrated subsystem level provides multiple benefits in the development process, which are summarized in Sect. 1.4.5. These serve as a basis for resolving the shortcomings in ride comfort, depicted in Sect. 1.2.2.

In a future prospect, there still exist several conditions concerning the application of the derivation process. When designing characteristic curves, like the vertical stiffness in Fig. 1.9, limitations in the flexibility of the realization are currently based on the experience of the developer. For instance, negative stiffness or discontinuities are designable while being practically not feasible. In this context specific objective boundary conditions for limitations and form of the characteristic curves are not given yet. Also while individual changes in parameters on subsystem level are purposeful for comprehension of effects, in a design process only a specific amount of difference between different degrees of freedom is practically applicable. An example is given by the difference between longitudinal and lateral stiffness of the suspension, both usually predominantly defined by bushings. While low stiffness in longitudinal and high stiffness in lateral direction are desired and designable in the subsystem approach, limits need to be defined for a maximum difference, as only specific configurations can practically be realized. Initially referencing on conditions on component level can be purposeful, where only a maximum difference in the stiffness between two axes of a bushing is designable. Still, the matter needs to be investigated in more detail as the composition of effects on subsystem level is higher than on component level.

Eventually, when the process is continuously defined and the frequency range is extended, also the modelling on subsystem level is to be modified. Therefore new subsystems need to be integrated or existing subsystems have to be detailed. For instance, an increase of the examined frequency range from 30 to 50 Hz would furthermore require the incorporation of effects due to compliance of the body. As natural frequencies are determined in an early phase of the development process, also a subsystem approach instead of finite element methods becomes possible. Appropriate models can eventually be defined using multi-body simulation with few degree of freedoms or modal models based on the given natural frequencies.

Still, it has to be ensured that a distinct separation of properties defining subsystem and component models is maintained. Otherwise an efficient derivation between different levels of detail of the vehicle is impeded.

References

Albrecht, M., Albers, A.: Einsatz Künstlicher Neuronaler Netze zur objektiven Beurteilung des Schwingungskomforts am Beispiel des automatisierten Anfahrens. In: Humanschwingungen, VDI-Berichte 1821. VDI Verlag GmbH, Düsseldorf (2004)

Angrick, C., Prokop, G., Knauer, P., Wagner, A.: Improved prediction of ride comfort characteristics by considering suspension friction in the automotive development process. In: chassis. tech plus – 6. Internationales Münchner Fahrwerk-Symposium, München (2015)

Bindauf, A., Angrick, C., Prokop, G.: Fahrwerkscharakterisierung an einem hochdynamischen Achsprüfstand. ATZ. **December**, 76–81 (2014)

Bock, T., Maurer, M., van Meel, F., Müller, T.: Vehicle in the Loop – Ein innovativer Ansatz zur Kopplung virtueller mit realer Erprobung. ATZ. **110**, 10–16 (2008)

Braess, H.-H., Seiffert, U.: Handbuch Kraftfahrzeugtechnik, 6th edn. Vieweg+Teubner, Wiesbaden (2011)

Cucuz, S.: Schwingempfindung von Pkw-Insassen: Auswirkung von stochastischen Unebenheiten und Einzelhindernissen der realen Fahrbahn. Dissertation, TU Braunschweig (1993)

Decker, M.: Zur Beurteilung der Querdynamik von Personenkraftwagen. Dissertation, TU München (2009)

Einsle, S., Fritzsche, C.: Utilization of objective tyre characteristics in the chassis development process. In: chassis.tech plus – 4. Internationales Münchner Fahrwerk-Symposium, München (2013)

Gillespie, T.: Fundamentals of Vehicle Dynamics, 1st edn. SAE International, Warrendale (1992)

Hab, G., Wagner, R.: Projektmanagement in der Automobilindustrie, 4th edn. Springer Gabler, Wiesbaden (2013)

Heißing, B., Brandl, H.-J.: Subjektive Beurteilung des Fahrverhaltens, 1st edn. Vogel, Würzburg (2002)

Heißing, B., Ersoy, M., Gies, S.: Fahrwerkhandbuch, 3rd edn. Springer-Vieweg, Wiesbaden (2011)

Hennecke, D.: Zur Bewertung des Schwingungskomforts von Pkw bei instationären Anregungen. Dissertation, Technische Universität Carolo-Wilhelmina zu Braunschweig (1994)

Holdmann, P., Köhn, P., Möller, B., Willems, R.: Suspension kinematics and compliance – measuring and simulation. SAE technical paper 980897 (1998). doi:10.4271/980897

Klingner, B.: Einfluss der Motorlagerung auf Schwingungskomfort und Geräuschanregung im Kraftfahrzeug. Dissertation, TU Braunschweig (1996)

Matschinsky, W.: Radführungen der Straßenfahrzeuge, 3rd edn. Springer, Berlin (2007)

Mitschke, M., Wallentowitz, H.: Dynamik der Kraftfahrzeuge, 5th edn. Springer Vieweg, Wiesbaden (2014)

Nakahara, J., Minakawa, M., Gipser, M., Wimmer, J.: A modelling approach to suspension friction. AutoTechnology. **1**(3), 54–56 (2001). doi:10.1007/BF03246609

Pacejka, H.: Tyre and Vehicle Dynamics, 2nd edn. Butterwort-Heinemann, Oxford (2006)

Rauh, J.: Virtual development of ride and handling characteristics for advanced passenger cars. Veh. Syst. Dyn. **40**(1–3), 135–155 (2003). doi:10.1076/vesd.40.1.135.15876

Schimmel, C.: Entwicklung eines fahrerbasierten Werkzeugs zur Objektivierung subjektiver Fahreindrücke. Dissertation, TU München (2010)

Seiffert, U., Rainer, G.: Virtuelle Produktentstehung für Fahrzeug und Antrieb im Kfz, 1st edn. Vieweg+Teubner, Wiesbaden (2008)

Stammen, K., Meywerk, M.: Anwendbarkeit künstlicher neuronaler Netze auf die Bewertung des Schwingungskomforts im Kraftfahrzeug – Eine Untersuchung im Rahmen der virtuellen Fahrzeugentwicklung. In: Humanschwingungen, VDI-Berichte 2002. VDI Verlag GmbH, Düsseldorf (2007)

Yabuta, K., Hidaka, K., Fukushima, N.: Effects of suspension friction on vehicle riding comfort. Veh. Syst. Dyn. **10**(2–3), 85–91 (1981)

Zegelaar, P.: The dynamic response of tyres to brake torque variations and road unevennesses. Dissertation, TU Delft (1998)

Chapter 2
Methods for Change Management in Automotive Release Processes

Christina Singer

Abstract The handling of changes in automotive release processes is a fundamental challenge of today's development projects. This chapter examines strategies for the identification of the effects of changes and evaluates concepts for the estimation of resulting retest effort. It is determined that there exists no approach that is applicable for large systems at vehicle level and that allows a reliable selection of all tests necessary to analyze the impact of the change. To solve this problem, two general concepts for test selection techniques are proposed. Inclusion-based approaches identify tests from the set of not executed tests whereas exclusion-based approaches eliminate tests from the set of performed tests. The two concepts are compared via receiver operating characteristic and cost estimation. Furthermore, the exclusion-based test selection is described in detail. It offers the opportunity to reduce the automotive release effort without drawbacks in test quality.

Keywords Test selection • Change management • Automotive release process • Inclusion • Exclusion • Development process • Validation • Retest

2.1 Introduction and Motivation

The number of software functions in vehicles is continuously rising due to increased market demands and improved availability of power electronics. Today's premium cars have up to 80 Electronic Control Units (ECUs), which are connected via multiple system busses and realize several thousand functions (Broy et al. 2011). This trend is accompanied by raised individualization, which results in a huge variety of models and configurations and therefore in a high complexity of automotive systems. Together with increased safety and reliability requirements, this leads to rising effort for testing and approving the total system. Thus, the release process has become a crucial element in the development process. Moreover, the high innovation and cost pressure in the automotive industry calls for shorter development cycles and causes fast changing platforms and system

C. Singer (✉)
Mozartstr. 8a, 91083 Baiersdorf, Germany
e-mail: christina_singer@gmx.net

H. Winner et al. (eds.), *Automotive Systems Engineering II*,
DOI 10.1007/978-3-319-61607-0_2

infrastructures. The rapid integration of new technologies as well as the efficient handling of changes are therefore essential competitive factors. Hence, the handling of changes in automotive release processes is a fundamental challenge of today's development projects (Broy 2006; Fürst 2010; Sundmark et al. 2011).

Changes occur due to functional extensions, functional changes, cost reduction activities, or adaptations to new hardware (Gustavsson 2010). To determine the effects of a change and to ensure that the safety of the system has not been affected by the modification, the release process has to be revised. The identification of the effects of the change and the calculation of the resulting retest effort is the fundamental challenge in this situation. Today, the selection of necessary test cases at vehicle level is mostly carried out manually by the test engineer. Therefore, the quality of the result depends primarily on the experience and expertise of the tester. Support in terms of a comprehensible and reproducible methodology that also facilitates legal certainty does not exist. Consequently, the entire test suite is retested in practice which leads to high expenses particularly in real vehicle driving maneuvers, or a subset of tests is selected on the basis of uncertain criteria which results in a residual risk (e.g. recall).

This chapter is mainly based on the results of Singer (2016). It examines strategies for the identification of the effects of changes and evaluates existing approaches for the determination of resulting retest effort. Furthermore, it proposes two general concepts for test selection: exclusion- and inclusion-based techniques. The two methods are compared via receiver operating characteristic (ROC) analysis and cost estimation. Furthermore, the exclusion-based concept is described in detail. It supports release decisions at vehicle level and avoids the disadvantages of the state-of-the-art technologies.

The chapter is structured as follows: Sect. 2.2 provides some basic information about automotive release processes and change management strategies. State-of-the-art methods for change propagation analysis and retest effort estimation are described in Sect. 2.3 and evaluated in Sect. 2.4. Section 2.5 illustrates the two general test selection approaches. The exclusion-based concept is presented in Sect. 2.6. Section 2.7 concludes the chapter.

2.2 Automotive Release Processes and Change Management

2.2.1 Introduction to Automotive Release Processes

The automotive development process for electronic systems is oriented on the V-model (Schäuffele and Zurawka 2003, p. 19). The V-model is a graphical representation of a system development cycle. It was introduced in 1992 to improve software development processes (Rausch and Broy 2008, p. 2) and summarizes the main steps in a development project (see Fig. 2.1).

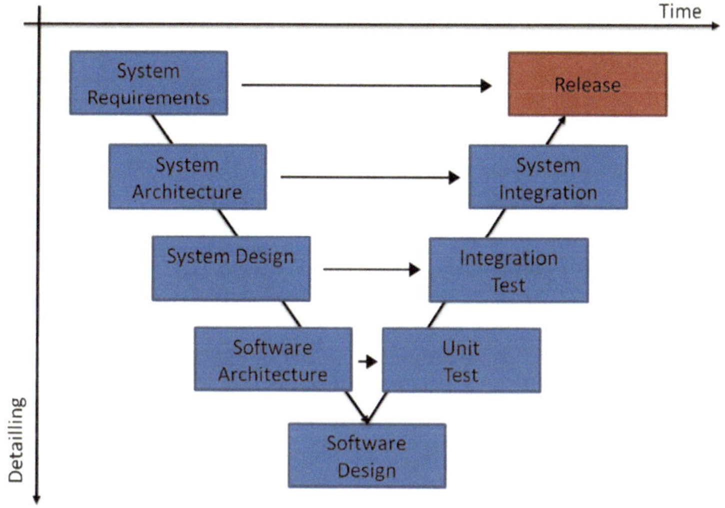

Fig. 2.1 V-Model (according to HTWK Leipzig 2014)

The left side of the "V" includes a top-down process that starts with the definition of system requirements and ends in the implementation of software elements. The right side of the "V" follows a bottom-up process and involves system integration and test activities. The release process is the final step of the V-model (see red box in Fig. 2.1). Here, the developed system is tested against its system requirements. Therefore, the release process bridges the gap between system development and operational use (Schäuffele and Zurawka 2003; Reif 2007).

A short release process facilitates a fast market launch but needs efficient integration and test processes. Therefore, the release is one of the most important elements of the development process (Sundmark et al. 2011).

2.2.2 Regulatory Framework for Automotive Release Processes

Because the release bridges the gap between system development and its operational use by the customer, it has to assure that the system fulfills all requirements. Therefore, the release decision is of high legal relevance. Product liability laws, for example the German Produkthaftungsgesetz (ProdHaftG), ensure that the manufacturer as well as the suppliers are held responsible for their products. Therefore, they have to make certain that their products are developed according to the present state-of-the-art. The state-of-the-art represents current laws, regulations and standards, as well as patents and publications (Reuter 2011).

Two kinds of requirements are taken into account in automotive release processes. On the one hand, there are requirements that focus on the properties and functionalities of vehicles. Examples for this category are vehicle certification laws, e.g. UN ECE R13 H (2015) or FMVSS 126 (2007) for automobile brake systems, or safety standards e.g. ISO 26262 (2011) for the functional safety of road vehicles. On the other hand, there are requirements that deal with the development process. Examples for this class are IATF 16949 (2016), which focuses on quality management in the automotive industry, or Automotive SPICE (2015), which includes software development processes.

2.2.3 Implementation of Automotive Release Processes

The release is the result of approval and acceptance tests for the developed and parameterized system. The tests are part of a formal process and include verification and validation activities (Sundmark et al. 2011; Reif 2007). Release tests are usually black-box tests, where the functionality of a system is examined without insight into its internal structure (Borgeest 2008). They consist of a great variety of different tests, which focus on diverse targets. Typical test categories for automotive controllers are for example (Borgeest 2008):

- Functional tests
- Robustness tests
- Recovery tests
- Benchmarks
- Configuration and compatibility tests
- Usability tests
- Security tests
- Endurance tests.

Release tests presume that the system is tested in its final environment under customer operating conditions to ensure that all requirements are fulfilled. Because of that, a huge amount of release tests is executed in real vehicles. Common test targets are driving dynamics, acoustics, reference and benchmark (e.g. motorsport magazine comparison tests), test of driver assistance systems, and drivability. In many cases, the assessment of driving characteristics is subjective (e.g. on a scale from 1 to 10). A complete objective evaluation of driving maneuvers is not state-of-the-art and therefore a major challenge for the release process (Sundmark et al. 2011; Düser 2010).

Another environment for release tests are hardware-in-the-loop-(HIL) tests. Here, the developed system, for example a brake system control, is tested in a simulated vehicle environment. High repeatability of driving maneuvers and the opportunity for test automation are advantages of the simulation approach compared to real vehicle tests. Moreover, the simple availability of different environmental conditions and diverse vehicle parameters is advantageous (Borgeest 2008).

Kvasnicka et al. (2006) and Mao et al. (2012) show examples for the application of HIL-simulation tools in brake system release processes. Furthermore, a CAE-based homologation approach for ESC systems is described by Holzmann et al. (2012). The quality of the HIL-tests depends significantly on the data (e.g. mathematical vehicle model, vehicle parameters), which was used to build the models. A high degree of detail usually requires a huge modeling effort (Langermann 2008). Furthermore, not all evaluation aspects (e.g. driving comfort, drivability, and usability) can be objectively measured. Therefore, the use of HIL-tests in release processes is limited.

2.2.4 *Sources of Change*

Changes occur regularly in the course of a development project. They arise due to functional extensions, functional changes, cost reduction activities, or adaptations to new hardware (Gustavsson 2010). Eckert et al. (2004) distinguish between initiated changes and emergent changes. Initiated changes arise from an outside source. Many of these changes are known at the beginning of the development process, e.g. customer demands, legal requirements for products or processes, innovations (new materials, components, software), problems with past designs. Some arise during a development project due to new customer requirements or recent innovations. Emergent changes are caused by the state of the design. They result from problems with the actual product. Problems arise during all stages of the development process and at all integration levels, e.g. in design, testing, prototyping, manufacturing, or use.

The later a change occurs in a development project, the more expensive is its implementation. There are two reasons for this. On the one hand, the processes become more time-critical because there is less time left for the handling of the change. Therefore, more resources are needed to deal with the change in time. On the other hand, the product is more integrated, so that the impact of the change is larger and more rework and retests are necessary (Eckert et al. 2004). Changes during the release process, which forms the last step before the product is produced and delivered to the customer, are therefore a fundamental challenge.

2.2.5 *Automotive Change Management*

The handling of changes is organized by change management processes. A change is therein treated as a small project within the overall vehicle development project (Borgeest 2008). Gustavsson (2010) describes the change management process as a five step action. First, the change is identified through a demand analysis. Thereon, the impacts of the change are determined. In the next step, solution alternatives are set up. Afterwards, the decision about the change is reached. Finally, the implementation and validation of the change take place.

Automotive change management is defined in Automotive SPICE (2015). The specified process aims at ensuring that changes are controlled, tracked, and monitored. Therefore, a change management plan is established, which contains the change management strategy of the company. Here, change management activities including identification, documentation, analysis, and implementation of changes are defined. The central element of the change management process is the so-called change request (CR), which is used to handle a single change. It defines the purpose of the change, identifies its effects for existing systems, and contains the responsibilities as well as the status and the criticality of the change.

Requirements for handling changes in the automotive industry also arise from ISO 26262 (2011), which contains automotive functional safety standards. ISO 26262 (Part 8, Clause°7, 2011) describes the required change management process. It includes following steps: change request, change analysis, change decision, and change implementation and documentation. The central element is the analysis of the effects of the change on the functional safety of the system. According to the results of this analysis, the products of the safety lifecycle are adjusted and the affected work products are revised. Moreover, a revalidation of the concerned parts of the product is required. Also, the release process has to be reevaluated.

2.2.6 Summary and Conclusion

There are four major challenges of automotive release processes. The first challenge emerges from the high system complexity of the vehicle overall system, which results from the huge number of individual systems (hardware and software) and the intense connectivity between those systems. An entire testing of all combinations of input parameters of functions is not feasible with economically justifiable effort (Nörenberg 2012, p. 26). Therefore, test case selection is an essential activity.

The second challenge is the vast test effort for automotive releases due to the high amount of vehicle variants and the fact that the major portion of tests is conducted in real vehicle environments. The expenditures for verification and validation activities form a huge amount of the overall development costs (Albers 2010). This includes costs for vehicles and components, test tracks, simulation environments, manpower etc. Hence, the reduction of release effort is highly recommended.

The third challenge is the short time for the release process. The approval cannot be carried out until the final hardware and software are available (Sundmark et al. 2011). Because of the short overall development cycle in the automotive industry, the time for test actions in the release process is limited to a few months. For that reason, a high efficiency of the release activities is required.

The forth challenge of automobile release processes are changes. According to statements from interviewees at Scania CV, changes cannot be completely avoided

(Sundmark et al. 2011). They are especially critical if they occur late in the release process, when there is not much time left for their implementation and the retest of the system. Thus, the handling of changes in automotive release processes is a critical success factor. The identification of the effects of changes and the determination of resulting retest effort is therefore an important research field.

2.3 Methods for Change Propagation and Retest

Changes in components or software functions can have direct or indirect impacts on other components or functions. There exist several approaches that deal with the identification of these effects and the determination of the resulting retest effort. A selection of them is described in the following section.

2.3.1 Change Propagation on Function and Component Level

Numerous concepts for the detection of the effects of changes on function and component level can be found in literature.

Several approaches use Design Structure Matrices (DSM). The idea goes back to Steward (1981) and Eppinger et al. (1994). A DSM is a methodology that facilitates the capture, modeling, analysis, and synthesis (within limits) of the interconnections of elements in system networks. The basis of the system model is a square matrix, where the system entities are mapped on the rows and columns of the matrix. The dependencies between the elements are represented by the cells of the matrix. Each cell displays a numerical or binary representation of the connection between the element in the column and the element in the row. The effects of changes can be determined by restructuring of the dependency matrix (DSM Web 2014; Clarkson et al. 2001).

Examples for change propagation methods that are based on this technique are the Component—Function Propagation Method (Flanagan et al. 2003) and the Change Propagation Method (CPM) (Clarkson et al. 2001; Keller et al. 2005; Jarratt et al. 2004). The Component—Function Propagation Method analyses the dependencies between system elements on a binary basis (0—no connection, 1—physical or functional connectivity), whereas the CPM calculates the risk that the element is affected by a change of the corresponding element for each relationship. The risk is described as a combination of the likelihood and the impact of the change. The assignment of concrete values for these risk numbers is performed by expert evaluation.

Cohen et al. (2000) present a related approach called Change Favorable Representation (C-FAR), which facilitates change propagation on attribute level. The

C-FAR product model consists of entities, attributes that describe the entities, and relations that represent the interactions between the entities. Entities are illustrated by vectors, whereas the dimension of the vector corresponds to the number of attributes of the entity. The different elements are related to each other by matrices, whereas the matrix components are called linkage values. They are able to show quantitatively how a change in one attribute will influence the other. Linkage values can be H (high linkage), M (medium linkage), or L (low linkage). To calculate the consequences of a change from a source entity to a target entity, an influence path composed of a series of vector and matrices multiplications is determined.

Raffaeli et al. (2007) perform the change propagation analysis on the basis of a model of the product architecture. The main functions of a system are represented as black-boxes, which are connected via input-output-relations. Relations between entities can be material, signal, or energy flows. Furthermore, a mapping of functions and physical components takes place, whereas components are modeled in detail (e.g. geometric properties, material, color etc.) and are linked by physical connections (e.g. welded joints) or conceptual interdependencies (e.g. position).

Another approach that uses a product model for determining the effects of changes is shown by Eckert et al. (2004). A product is represented by three types of elements: direct parameters, functional parameters, and behavioral parameters. Physical components are described by direct parameters (e.g. geometry, material, mass, power). Functions result from interactions between direct parameters. They are divided into desired parameters and side effect parameters (e.g. noise, vibration, EMI). The behavior of the product can then be determined from the interactions between the functions. The methodology facilitates the identification of the effects of changes in direct parameters on the behavior of the system.

The illustrated concepts for change propagation on function and component level have in common that they determine the effects of a change on the basis of a simplified system model. Some use mathematical representations such as matrices or vectors, whereas others build complex product models with detailed insights on product properties, functions, or behavior. As an advantage, all described methods are generally applicable to automotive systems at vehicle level. The results of the change propagation analysis are otherwise highly dependent on the quality of the underlying system model. Furthermore, the effort for the creation of the system model gets very high for huge systems with lots of interconnections.

2.3.2 Change Impact Analysis on Software Level

Numerous concepts for the determination of the effects of changes exist in software engineering literature. They are called Change Impact Analysis (CIA). CIAs aim at identifying the parts of a software system that are affected by a change. Based on a number of defined changes ("change set"), they estimate the software elements that are influenced by these changes and need additional modification ("impact set") (Yazdanshenas and Moonen 2012).

Lehnert (2011) presents an overview of CIA methods, which analyzes 150 different literature sources. He distinguishes three scopes for CIAs.

Methods of the first type investigate the influence of changes by drawing conclusions from source code. They analyze inheritance relations, method-call behavior, and other dependencies between program entities. Static approaches evaluate call graphs, slices, and other representations of source code whereas dynamic and online concepts analyze execution traces.

The second application area of CIA is formal models. They can be further composed into architectural and requirements models. Architectural models are for example UML (Unified Modeling Language, see OMG 2014) component diagrams that illustrate systems, sub-systems, components, and classes of a software system. UML models allow the determination of the effects of changes on a more abstract level than source code. If requirements are represented in a formal modeling language, they can also be analyzed for the assessment of change impacts.

The third scope of CIAs form miscellaneous artifacts. These can be documentation, bug trackers, or configuration files. Combinations of these types are also possible.

There exist a lot of concepts for the identification of change impacts in software engineering. Some methods need a specific information basis, e.g. source code. In automotive release processes, this information is not always available. For example, the vehicle manufacturer usually has no access to the source code of a brake system controller. Other methods use a less specific information basis, for example UML models. These concepts are generally applicable on systems at vehicle level. Otherwise, the effort for establishing and maintaining these models is high.

2.3.3 Regression Test Selection Techniques

Another concept for reducing release effort in the case of changes is the regression test. Regression tests aim at testing an already evaluated object again after its modification. They intend to prove that the implementation of the change has not led to further defects. Studies estimate that regression tests form about 80% of the overall test effort (Kaner 1997). The test effort can be reduced if just a selection of all test cases is reevaluated (Kim et al. 2005). Therefore, regression test selection (RTS) techniques are applied. Studies (Leung and White 2005; Rothermel et al. 2002; Khan et al. 2009) prove the effectiveness of this approach.

RTS methods use different information as a basis for the determination of the retest effort. Many concepts use representations of source code to identify the parts of the software that are affected by the change and assign test cases to these areas. Examples for this approach can be found by Vokolos and Frankl (1997), Rothermel and Harrold (1998), or Gallagher et al. (2007). Component-based RTS techniques are based on software elements for which source code is not available. They analyze the interfaces of software components or use call graphs as a basis for the impact analysis. Zheng et al. (2005) show an example of this method. Other concepts (Zhao et al. 2002; Muccini et al. 2006; Briand et al. 2009) utilize architecture models (e.g. UML-diagrams) that are connected with test cases to determine the effect of changes and to calculate the retest effort. Requirements are another source of information for RTS techniques. Gorthi et al. (2008) show a specification-based approach that is based on UML activity diagrams. Another example for this concept is Chittimalli and Harrold's Requirement RTS (2008), which links requirements with code and selects test cases by analyzing test traces.

Caliebe et al. (2012) and Nörenberg (2012) have transferred the concept of regression tests from software applications to embedded systems. Caliebe et al.'s regression test methodology uses a black-box system model called Component-Dependency-Model (CDM). It is derived from the system architecture (e.g. AUTOSAR architecture, see AUTOSAR 2014) and the corresponding system requirements and transformed into a graph structure to perform path analysis. Via graph algorithms, the effects of changes can be calculated and the necessary retest effort can be determined. Nörenberg (2012) describes a specification-based approach for regression tests that uses the concept of "Funktionsorientierung (FO)" (function orientation) from Daimler as a basis for the analysis of the change impact.

All described regression test selection techniques need a detailed representation of the analyzed system for the determination of the impact of the change. Some regression test selection techniques are very specialized, because they need a particular type of source code in a defined programming language. Other methods use a less specific information basis as for example UML models. Those concepts are generally applicable to systems on vehicle level.

2.3.4 Design of Experiments

Design of Experiments (DoE) offers another alternative to reduce test effort in release processes. The aim of this statistical test planning approach is the realization of desired experimental results with minimum test effort. By simultaneously changing several factors at a time, the influences of the different factors on a certain parameter or parameter group can be determined (Borgeest 2008).

Ungermann (2009) presents an approach that uses DoE to reduce the sample size of reliability tests in automotive release processes. Furthermore, the concept facilitates the systematic determination of necessary retest effort in cases of late design changes. The basis for the test planning forms an analysis of the component-specific failure behavior on different test tracks. Moreover, complexity classes of components and the degree of maturity of the development project are taken into account. By integration of information about the customer use, a model-specific planning standard is derived.

Burgdorf (2010) also uses DoE for the estimation of test effort in release processes of automotive E/E (Electric/Electronic) systems. On the basis of a prediction model for future power circuit configurations, representative vehicle configurations for release tests are identified. The concept is augmented by a customer-relevant risk definition that is used to calculate the statistical risk reduction, which can be achieved by different vehicle test configurations. A test plan is calculated by iterative optimization.

DoE approaches allow a targeted selection of test cases. As an advantage, they do not require considerable system knowledge, because they build on statistical data. They enable the tester to set particular focus on failure prone or risk afflicted configurations. Adversely, DoE approaches need a huge data basis that has to be available. Furthermore, the statistical nature of the concept may lead to test deficits if interactions of the change are statistically low.

2.3.5 Summary

The described methods for change propagation analysis and test selection can be distinguished based on their underlying information basis (see Fig. 2.2).

The first group uses software source code. Examples for this category are software CIAs. They allow a detailed analysis of the impact of a change, combined with a reliable selection of assigned test cases. In case of large, highly connected systems, the application of this approach is limited because of high computation effort. Also, this concept is restricted to special test objects (e.g. source code in C++) and therefore is not transferable to systems at vehicle level.

The second group builds on system models. This class can be further divided into mathematical system descriptions (e.g. matrices or vector representations as for example CPM), architecture models (e.g. UML models), product models (e.g. CDM), or specification models (e.g. FO). All methods of this kind are in principal applicable to systems at vehicle level. They permit the determination of the effects of changes on software level as well as on physical level. As a

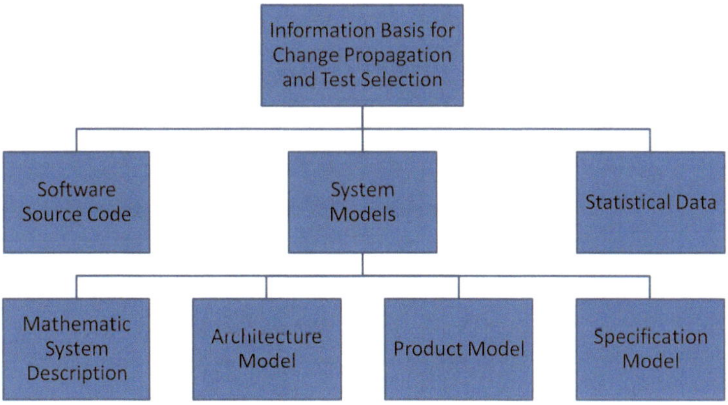

Fig. 2.2 Classification of the information basis for change propagation and test selection methods

disadvantage, the effort for the creation, maintenance, and administration of the models is high, especially for huge models with high interconnectivity. Partially, this effort can be reduced by automatic model generators. The result of the impact analysis depends highly on the accuracy of the model. Any interactions that are not modeled lead to deficits in test coverage.

The last group utilizes statistical data (e.g. DoE). These approaches enable a targeted selection of test cases without requiring considerable system knowledge. Otherwise, interactions between components or functions that are statistically unlikely are neglected, so that the test selection is not safe.

2.4 Evaluation of State-of-the-Art Methods

2.4.1 Retest Situations in Release Processes

The test situation in the case of changes in the release process can be described mathematically with set theory (see Fig. 2.3).

N represents the "true" set of total necessary tests to prove that a system is safe and can be delivered to the customer. The number of elements of *N*, $|N|$, is unknown.

T illustrates the total set of executed tests in the release process. The number of elements of *T*,$|T|$, is known. It corresponds to the sum of tests that are executed according to the state-of-technology. For example, for the release of brake systems, *T* responds to the total amount of driving maneuvers that are carried out by manufacturer and supplier during the release process.

In the case of changes, it is normally not necessary to do a complete retest, because the impact of the change is limited to some extent. N', $N' \in N$, represents

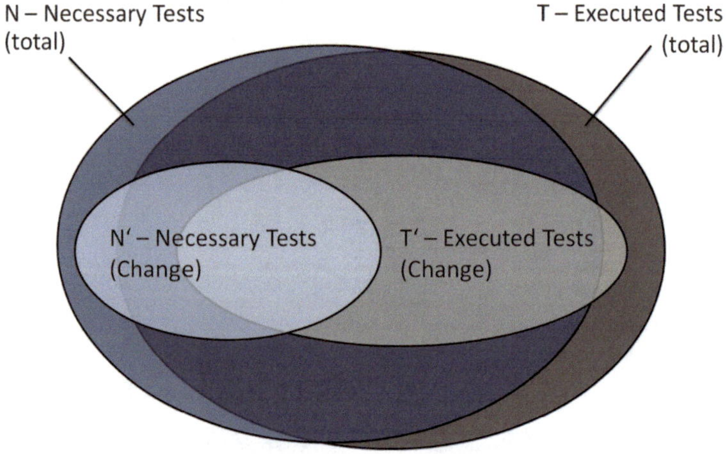

Fig. 2.3 Test situation in the case of changes in the release process

Table 2.1 Result matrix for test selection techniques

	Necessary tests $t_p + f_n = \lvert N' \rvert$	Obsolete tests $f_p + t_n = \lvert T \setminus N' \rvert$
Executed tests $t_p + f_p = \lvert T' \rvert$	True positive $t_p = \lvert T' \cap N' \rvert$	False positive $f_p = \lvert T' \setminus N' \rvert$
Not executed tests $f_n + t_n = \lvert T \setminus T' \rvert$	False negative $f_n = \lvert N' \setminus T' \rvert$	True negative $t_n = \lvert (T \setminus T') \setminus N' \rvert$

the "true" set of tests that is necessary to check all direct and indirect effects of the change, whereas T', $T' \in T$, describes the executed tests. This set is determined by expert evaluation or selection technique.

The result of a test case selection (manually or automated) can be distinguished in four categories (see Table 2.1). The first category includes those tests that are executed and necessary (true positive, t_p). The second category (false positive, f_p) incorporates those elements that are executed but obsolete. These test cases show the cost reduction potential of the selection technique. Tests that are not executed and obsolete form the third category (true negative, t_n). The last category (false negative, f_n) contains elements that are not executed but necessary. These tests mark the risk of the selection technique.

2.4.2 Evaluation Criteria for Test Selection Techniques (TST)

According to Rothermel and Harrold (1996), following criteria can be used to evaluate test selection techniques.

Inclusiveness

Inclusiveness describes the true-positive rate, i.e. the fraction of the selected necessary tests on the total number of necessary tests. It measures the sensitivity of a selection technique. A selection technique is called safe if the inclusiveness equals 1.

$$P_{tp} = \frac{t_p}{t_p + f_n} = \frac{|T' \cap N'|}{|N'|} \tag{2.1}$$

Precision

Precision illustrates the positive prediction value, i.e. the fraction of the necessary tests on the total number of selected tests. Precision measures the accuracy of the selection method.

$$P_{\mathrm{p}} = \frac{t_{\mathrm{p}}}{t_{\mathrm{p}} + f_{\mathrm{p}}} = \frac{|T' \cap N'|}{|T'|} \tag{2.2}$$

Efficiency
Efficiency evaluates the effort that results from the selection of test cases. Rothermel and Harrold distinguish time and space efficiency. A selection technique is more time efficient if the cost of selecting *T'* is less than the cost of running the tests in *T-T'*.

$$Cost_{\mathrm{Selection}}(T') < Cost_{\mathrm{Running}}(T - T') \tag{2.3}$$

Space efficiency assesses the technical requirements of the selection technique. Here, the needed information basis, which has to be determined, maintained, and documented, as well as the initial effort for installing the test selection concept are evaluated.

Generality
Generality demonstrates the ability of the selection technique to be applicable in a broad variety of situations. Factors that influence generality are the adaptability of the selection methodology to diverse systems and different kinds of changes.

2.4.3 Evaluation Results

In the following, the state-of-the-art change propagation and test selection methods discussed in Sect. 2.3.5 are evaluated on the basis of the criteria by Rothermel and Harrold (1996).

Inclusiveness
Software source code based TST offer a high inclusiveness. They analyze the effects of changes very detailed on the basis of the implemented code (e.g. control flow graphs). Therefore, they select test cases very reliably. Many concepts are therefore safe (see Rothermel and Harrold 1996). System model based TST use system abstractions as a groundwork for the determination of the impact of changes. Thus, their inclusiveness depends on the quality of the underlying system model. Each not modeled interaction leads to a reduction in test inclusiveness. For that reason, they are generally not safe. The inclusiveness of statistical data based TST is low because interactions between components or functions that are statistically unlikely are neglected.

Precision
The precision of software source code based TST depends on the concrete algorithm. According to Rothermel and Harrold (1996), no technique is 100% precise.

System model based TST select only tests that can be associated to the change by a modeled dependency. Hence, their precision is high. This also applies to statistical data based TST as they choose only tests that are statistically relevant.

Efficiency
All methods are time efficient because their algorithms are automated. The space efficiency of the concepts differs. Whereas software source code based TST are very space efficient because the underlying software is directly available, system abstractions for system model based TST and information for statistical data based TST have to be created and maintained (sometimes even manually), which results in high costs.

Generality
Software source code based TST are restricted to special test objects (e.g. source code in C++ or Java). Consequently, their generality is low. The other two methods are not designed for a concrete test object. So, they are applicable on diverse kinds of systems and different kinds of changes. Therefore, their generality is high.

The results for the evaluation of the discussed state-of-the-art change propagation and test selection methods are summarized in Table 2.2.

A TST that is applicable for the determination of retest effort in the case of changes in the release process has to fulfill several requirements. Due to the fact that the release decision is of legal relevance (see Sect. 2.2.2), the results of the selection

Table 2.2 Evaluation results for state-of-the-art change propagation and test selection methods

	Software source code based TST	System model based TST	Statistical data based TST
Inclusiveness	High Many concepts are safe.	Low/High Depends on the quality of the underlying system model, generally not safe.	Low Just statistically relevant interactions are considered.
Precision	Low/High Depends on the concrete algorithm.	High Only necessary test cases are selected.	High Only necessary test cases are selected.
Time efficiency	High Algorithms are automated.	High Most methods are automated; some need manual input from experts.	High Algorithms are automated.
Space efficiency	High Process is automated; software code is directly available.	Low System models have to be created and maintained; most of them have to be edited manually.	Low/High Depends on the required data, which has to be collected and maintained.
Generality	Low Restricted to software systems.	High Applicable on diverse kinds of systems, including software and hardware, and different kinds of changes.	High Applicable on diverse kinds of systems, including software and hardware, and different kinds of changes.

technique have to be safe, i.e. inclusiveness equal to 1 is targeted. Because of the large release effort in general (see Sect. 2.2.6), the selection technique should also select as few test cases as possible. Therefore, its precision should be high. To be of practical use in automotive companies, the selection method should be efficient in terms of time and space. Therefore, it is required that it uses already existing information and is capable of being integrated into existing processes. Furthermore, the release process validates systems on vehicle level. Therefore, it is obligatory that the selection technique has a high generality in terms of adaptability on diverse kinds of systems and different kinds of changes.

The evaluation results for the state-of-the-art change propagation and retest methods show that there exists no approach that allows a safe (in terms of high inclusiveness) selection of tests in the case of changes in the release process that is applicable on systems at vehicle level. The research question therefore is: Does a test selection technique exist that solves this dilemma? How this question can be answered is presented in the next section.

2.5 General Approaches for Test Selection Techniques

There exist two generic concepts for the selection of tests in the case of changes in the release process. They are described and compared in the following.

2.5.1 Inclusion-Based Test Selection

An inclusion-based test selection concept asks the question: Which test cases have to be executed? Hence, the target of this approach is the identification of true positives (t_p). The initial set of selected tests for the evaluation of the effects of the change is empty if an inclusion-based test selection approach is performed:

$$T'_{\text{initial}} = \{\ \} \rightarrow t_n + f_n = |T| \text{ or rather } t_p + f_p = 0.$$

On the basis of an impact analysis that determines the effects of the change, necessary test cases (t_p) are identified out of T and transformed into the set of executed tests T'. Therefore, as many false negative tests (f_n) as possible have to be transferred into true positive tests (t_p) (see Fig. 2.4).

2.5.2 Exclusion-Based Test Selection

An exclusion-based test case selection concept asks the question: Which test cases do **not** have to be executed? Hence, the target of this approach is the identification

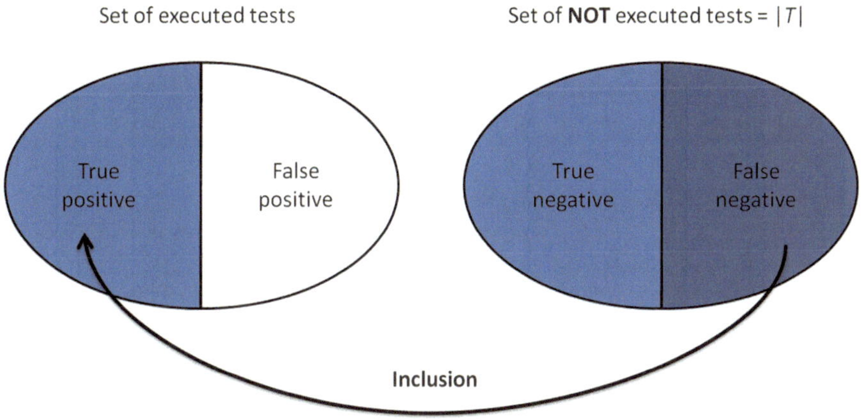

Fig. 2.4 Inclusion-based test selection technique

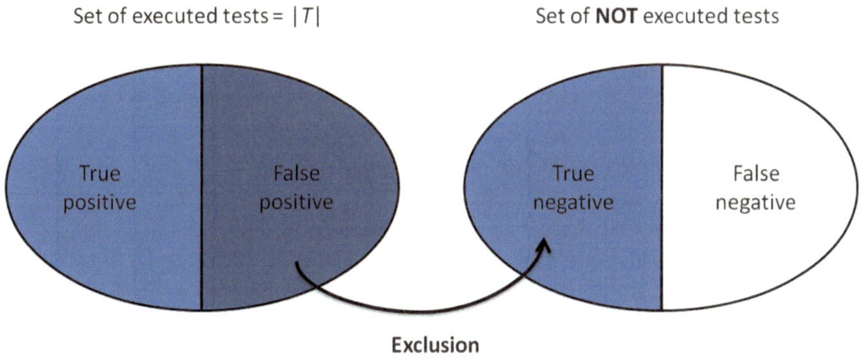

Fig. 2.5 Exclusion-based test selection technique

of true negatives (t_n). If an exclusion-based test case selection approach is carried out, the initial set of selected test cases for the evaluation of the effects of the change equals the whole test suite T.

$$T'_{\text{inital}} = T \rightarrow t_p + f_p = |T| \text{ or rather } t_n + f_n = 0.$$

Therefore, as much as many positive tests (f_p) as possible have to be transformed into true negative tests (t_n) (see Fig. 2.5).

2.5.3 Performance Comparison

The two approaches have different characteristics. The effects of these characteristics on the selection quality and the selection costs are presented in the following sections.

Comparison of Selection Quality

The selection quality of a test technique can be examined in terms of inclusiveness and error rate. A graphical approach for performing this analysis is the receiver operating characteristic (ROC). This method was originally used in signal detection theory to differentiate known signals from random noise. Today, the technique is applied in a variety of areas including psychology, radiology, finance, social science, and machine learning (Tan 2009). ROC represents a two-dimensional graph, where the true positive rate (*TPR*) (e.g. $t_p/(t_p + f_n)$) is plotted against the false positive rate (*FPR*) (e.g. $f_p/(f_p + t_n)$). Thereby, the trade-off between the successful detection of positive examples and the false classification of negative examples can be examined. (Tan 2009).

Figure 2.6 shows the ROC curves obtained by the two test selection approaches assuming an ideal selection quality (SQ) of 1 ($SQ = (t_p + t_n)/|T|$).

Inclusion-based methods identify false negative examples from the set of not executed tests and transform them into true positives. As the preliminary set of true positives equals zero, the *TPR* is initially zero. By adding true positives to the set of executed tests, the *TPR* increases until it reaches 1, when all true positives are detected. As inclusion-based techniques select only tests that are necessary, all obsolete tests remain in the set of not executed tests. False positive results are therefore avoided. Thus, the *FPR* is always zero.

Exclusion-based approaches identify false positive examples from the set of executed tests and convert them into true negatives. The *FPR* is therefore initially

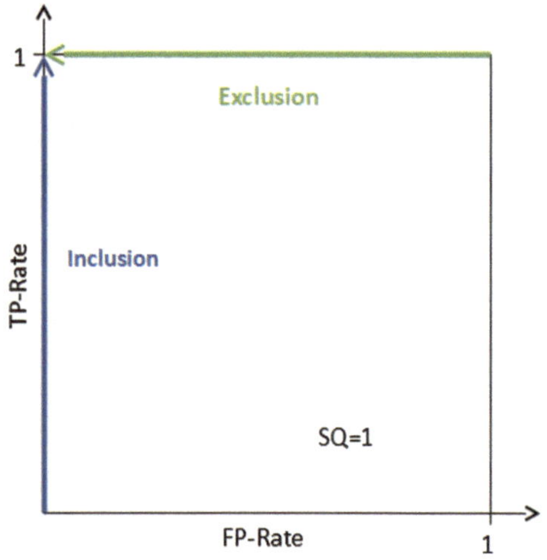

Fig. 2.6 Performance comparison of test selection approaches using ROC curves

1. By adding true negatives to the set of not executed tests, the *FPR* is reduced until it gets zero, when all true negatives are obtained. The *TPR* always equals 1 because all necessary tests stay in the set of executed tests. False negative results are therefore prevented, which shows that this method is always safe in terms of inclusiveness (see Sect. 2.4.2).

The ROC curves show that both techniques reach the same point in the ROC diagram ($TPR = 1$, $FPR = 0$) when their selection quality is ideal. Their starting points differ considerably. Whereas inclusion-based methods launch from ($TPR = 0$, $FPR = 0$), exclusion-based approaches begin at ($TPR = 1, FPR = 1$). It is therefore necessary to analyze if this difference has an influence on the costs of the two generic concepts.

Comparison of Expected Selection Costs

The performance of the two generic test selection approaches can also be determined by an examination of the methods' costs. Drummond and Holte (2000) propose a cost model for classifiers in machine learning which can be adapted for this problem. They suppose that all costs are finite and always strictly greater than zero. They further assume that the cost of correctly classifying an example is always less than the cost of misclassifying it. The best possible test selection technique therefore classifies every test correctly and has an expected cost of zero. The expected costs, *EC*, for a selection technique are calculated as follows:

$$EC = EC1 + EC2 = FNR \cdot p(+) \cdot C(-|+) + FPR \cdot p(-) \cdot C(+|-). \qquad (2.4)$$

FNR represents the false negative rate ($f_n/(t_p + f_n)$), whereas *FPR* illustrates the false positive rate ($f_p/(f_p + t_n)$). $p(+)$ is defined as the probability of a test being in the positive set, e.g. the set of executed tests, and $p(-) = 1\text{-}p(+)$ is the probability of a test being in the negative set, e.g. the set of not executed tests. The probabilities $p(-)$ and $p(+)$ are unknown. Their magnitude depends on the size of effect of the examined change. As the size of the effect can vary highly, a test selection technique has to be applicable for all distributions of $p(-)$ and $p(+)$. $C(-|+)$ and $C(+|-)$ allow a different weighting of misclassification costs. $C(-|+)$ characterizes the costs of misclassifying positive examples (f_n), whereas $C(+|\text{-})$ corresponds to the costs of misclassifying negative examples (f_p).

The maximum expected costs, max *EC*, occur, when all tests are incorrectly classified, i.e. when $FPR = 1$ and $FNR = 1$:

$$\max EC = p(+) \cdot C(-|+) + p(-) \cdot C(+|-). \qquad (2.5)$$

The normalized expected costs, *NEC*, are calculated by dividing the expected costs, *EC* (Eq. 2.4), by the maximum possible expected costs, max *EC* (Eq. 2.5):

$$NEC = \frac{FNR \cdot p(+) \cdot C(-|+) + FPR \cdot p(-) \cdot C(+|-)}{p(+) \cdot C(-|+) + p(-) \cdot C(+|-)}. \tag{2.6}$$

In the first evaluation step no weighting of misclassification costs is included, e.g. $C(-|+) = C(+|-)$. Under this condition, the normalized expected costs, *NEC* (Eq. 2.6), can be simplified as follows:

$$NEC_{\text{simp}} = FNR \cdot p(+) + FPR \cdot p(-) \tag{2.7}$$

Under the assumption of equal misclassification costs, the expected costs of the inclusion-based test selection only consist of cost parts that originate from false negative tests (f_n), as false positive results (f_p) are prevented by this strategy. Therefore, only the first cost summand has to be considered. The expected costs of exclusion-based approaches only consist of cost portions arising from false positive tests (f_p), as false negative results (f_n) are avoided by this method. Consequently, only the second cost summand has to be regarded here. If, in addition, an ideal selection quality is supposed, all tests are classified correctly at the end of the selection process. In this case, no selection costs arise at all and both generic concepts can be evaluated as equal.

If an ideal selection quality cannot be realized in practice, the expected selection costs depend on the achievable error rates FNR (for inclusion-based methods) and FPR (for exclusion-based approaches). Additionally, the different weightings of the misclassification costs $C(-|+)$ and $C(+|-)$ have to be taken into account. If the costs of false negative tests ($C(-|+)$) are higher than the costs of false positive results ($C(+|-)$), the scope of application of the exclusion-based test selection is increased. This is because exclusion-based methods avoid false negative results. Hence, only the less expensive costs of false positive classifications arise. If the costs of false positive tests ($C(+|-)$) are higher than the costs of false negative results ($C(-|+)$), the inclusion-based approach is preferred accordingly.

If the test selection regards changes at vehicle level, it can be assumed that the costs of incorrect classifications differ. On the one hand, false positive results lead to tests that are executed even if they are not required to evaluate the change. This causes expenses of about 10^{3-4} euros for each test that is categorized wrongly. False negative results on the other hand induce tests which are not carried out, even if they are necessary to assess the consequences of a change. This does not have to cause problems inevitably. However, it may happen that the lack of test coverage provokes errors not being found. In this particular case, the false negative classification may lead to interferences in the functionality of the vehicle or—in the worst case—to deficits in passenger safety. Then, the costs of the incorrect categorization may become huge. For example, a vehicle recall causes expenses of about 10^{6-8} euros.

As the two generic test selection approaches are examined under the assumption of an ideal selection quality so far, the differences in the costs of incorrect classifications are irrelevant, as with both methods all tests are classified rightly at the end of the selection process. If the premise mentioned above is not viable in practice, false negative results have to be avoided coercively under the point of view of expected selection costs. To use the inclusion-based test selection, it is then necessary to prove that the underlying model for the estimation of the effects of the change can describe the interactions of the change in a complete and correct way. Only in this case, false negative classifications can be eliminated.

Summary and Discussion

Table 2.3 summarizes the results of the performance comparison of inclusion- and exclusion-based test selection methods.

Inclusion-based methods are always exact. Otherwise, they are not necessarily inclusive. Exclusion-based techniques produce safe results, but they are not automatically error-free in terms of false positives. The expected costs of the two approaches are of different origin. The costs of an inclusion-based test selection only consist of cost parts that originate from false negative tests, whereas the costs of exclusion-based methods are composed of cost portions arising from false positive tests. If an ideal selection quality is supposed, all tests are classified correctly at the end of the selection process. In this case, no selection costs occur at all and both generic concepts can be evaluated as equal. Thus, no method can be preferred. It is therefore necessary to examine if the assumption of equal misclassification costs can be kept up when the methods get more concrete. Consequently, further detailing of the two concepts is required. This will allow a more detailed analysis of achievable selection performance and costs. In the following, a concept for an exclusion-based test selection technique is described.

Table 2.3 Comparison of inclusion and exclusion

	Inclusion	Exclusion
Inclusiveness	The inclusiveness of the inclusion-based approach is initially zero, as no true positives (t_p) are in the set of executed tests at first. A high inclusiveness is only achieved if all false negatives (f_n) are identified and transformed into true positives (t_p).	Exclusion-based methods are always safe in terms of inclusiveness. All necessary tests stay in the set of executed tests T‘. False negative results (f_n) are therefore avoided, i.e. $f_n = 0$.
Error rate	Inclusion-based methods are always exact. All obsolete tests remain in the set of not executed tests (T-T‘). False positive results (f_p) are therefore avoided, i.e. $f_p = 0$.	The error rate of the exclusion-based test selection is initially 1, as no true negatives (t_n) are in the set of not executed test (T-T‘) at first. An error rate of zero is only reached if all false positive tests (f_p) are identified and transformed into not executed tests (t_n-tests).
Expected costs	The expected costs of an inclusion-based test selection only consist of cost parts that originate from false negative tests (f_n) as false positive results (f_p) are prevented by this strategy.	The expected costs of exclusion-based approaches only consist of cost portions that arise from false positive tests (f_p) as false negative results (f_n) are avoided by this method.

2.6 Example: Exclusion-Based Test Selection Technique

Exclusion of test cases can be reached by a process of elimination. The state-of-technology marks the starting point for the exclusion. Therefore, the test set T contains all driving maneuvers which are carried out by the manufacturer and the suppliers during the release process. In the first step, only real vehicle tests are considered by the test selection technique due to the fact that they cause the main portion of costs in the release process.

A test t_i can be removed from the set of executed tests T and transformed into the set of excluded tests T_{ex} if evidence exists that confirms that the test t_i is not necessary. For that reason, the result $R(t_i)$ of the test t_i is observed. The result $R(t_i)$ evaluates the outcome of the test t_i, i.e. the behavior actually produced when the object is tested under the specified conditions of the test t_i (Test Glossary 2014). Test results can be determined on three different measurement levels (Kohn 2005, pp. 13–15). On a nominal scale, test results can be classified into different categories, i.e. passed or failed. On an ordinal scale, test results can be arranged in a ranking order. An example for this category is the ATZ scale (Aigner 1982) which is used to assess vehicle characteristics (i.e. agility, stability, comfort) on a rating between 1 and 10. On the third level, test results can be measured on a metric or cardinal scale. Examples for this class are brake distance or fuel consumption. To prove that a test is not necessary, the impact of the change C on the result R of the test t_i is analyzed. The effect of the change C can be sorted into different categories: First, the test results can be equal, i.e. $R(t_i,C) = R(t_i)$. Second, the results can be less or equal ($\leq$) or greater or equal ($\geq$), i.e. $R(t_i,C) \leq R(t_i)$ or $R(t_i,C) \geq R(t_i)$. Third, the results can be definitely unequal, i.e. $R(t_i,C) \neq R(t_i)$. According to this differentiation, a test case can be excluded from the set of executed tests, when it can be proved that the test result $R(t_i,C)$ is equal or better than the test result $R(t_i)$. Three exclusion arguments can be considered for this proof.

Exclusion Argument I: Result Neutrality

A test t_i ($t_i \in T$) can be excluded from the set of executed tests T, if a formal evidence exists, that demonstrates, that a change C has no impact on the result R_i of t_i.

$$\text{EA I}: \qquad R(t_i, C) \stackrel{!}{=} R(t_i) \rightarrow t_i \in T_{ex,1}$$

A change C has no impact on a test, when the test result $R(t_i, C)$ lies within the test reproducibility margin of the test result $R(t_i)$. To prove the result neutrality exclusion argument, formal evidence is required that the test is not affected by the change. In case of a modification of a software function, such evidence exists if it can be confirmed, for example by code review or formal verification, that the function f—within its range of variable parameters—can never be activated in the considered test scenario.

Exclusion Argument II: Single Result Equality
A test t_i ($t_i \in T$) can be excluded from the set of executed tests T, if another test t_j (($t_j \in T$) exists, whose result R_j shows the impact of a change C at least as well as the result R_i of t_i.

$$\text{EA II}: \quad R(t_i, C) \leq R(t_j, C) \rightarrow t_i \in T_{ex,2}$$

Exclusion Argument III: Multiple Result Equality
A test t_i (($t_i \in T$) can be excluded from the set of executed tests T, if a combination of tests $t_j \ldots t_n$ ($t_j \ldots t_n \in T$) exists, whose results $R_j \ldots R_n$ show the impact of a change C at least as well as the result R_i of t_i.

$$\text{EA III}: \quad R(t_i, C) \leq \sum_{k=j}^{n} R(t_k, C) \rightarrow t_i \in T_{ex,3}$$

As exclusion argument II is a special embodiment of exclusion argument III, both cases can be combined for further considerations. The result equality exclusion argument marks a test case t_i as redundant when the results of a combination of tests in the executed test set show the impact of the change at least as well as the result of t_i. For example, this can be verified when the test targets of a test t_i are covered by the remaining tests T—$\{t_i\}$. Here, the application of optimal test planning methods is also possible. Tests which are used as reference for the exclusion of a test t_i by the result equality argument, are not allowed to be further excluded. They have to stay in the set of executed tests. Hence, they experience an inclusion.

The resulting concept for the exclusion-based test selection technique is presented in Fig. 2.7.

Input for the test selection technique is an initial test suite T and a change C that has to be verified. In the first step, exclusion argument I (EA_I) (result neutrality) is evaluated. Hence, all test cases which are verifiable not affected by the change are eliminated and transformed into the set of excluded test $T_{ex,1}$. Thereby, the test suite is reduced. The remaining tests $T_{red,1}$ are analyzed in the second step. Here, exclusion argument III (EA_{III}) (result equality) is considered. Consequently, all

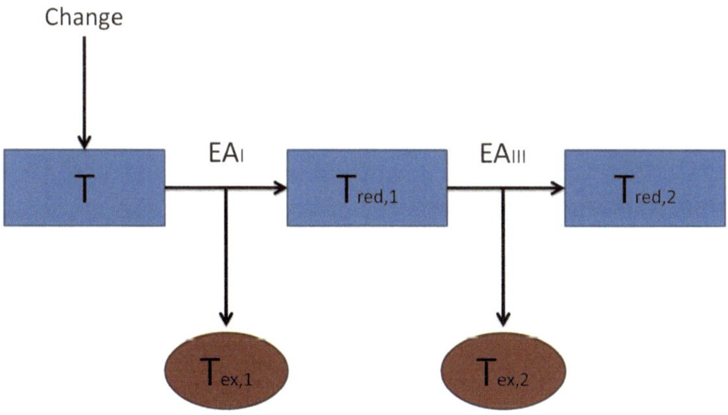

Fig. 2.7 Concept for exclusion-based test selection technique

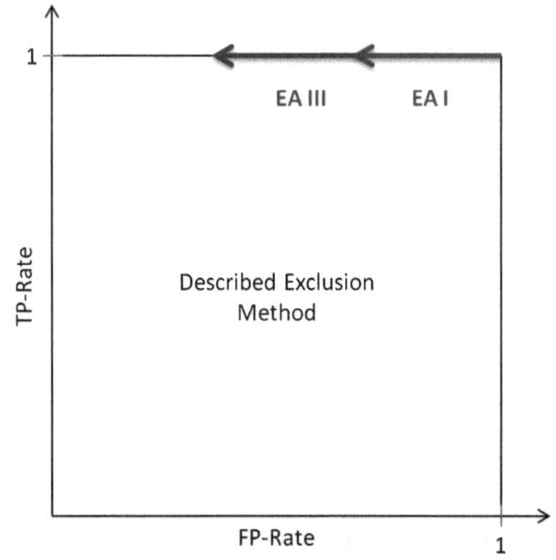

Fig. 2.8 Result of exclusion-based test selection technique

tests are removed from the set of executed tests and transferred to the set of excluded tests $T_{ex,2}$ whose results concerning the change can be accomplished by running the remaining tests $T_{red,2}$.

The result of the described exclusion-based test selection technique is shown in Fig. 2.8. The two exclusion arguments offer a large test reduction potential as they eliminate tests from the set of executed tests in a simply implementable way. Thereby, the false positive rate is decreased in two steps and the release test expenses can be shortened compared to the current state-of-technology. As it will not always be possible to verify formally, that a test is not affected by a change, the false positive rate will not be reduced entirely.

Simultaneously, the exclusion-based test selection technique preserves a high true positive rate. The resulting test suite $T_{red,2}$ is always safe in terms of inclusiveness. That's because only obsolete tests are excluded, so that all necessary tests remain in the set of executed tests. Hence, false negative results are prevented. Therefore, its application in legally relevant automotive release processes is promoted. Thereby, an efficient and legally secured handling of changes in the release process of automobile systems is facilitated.

2.7 Summary and Outlook

This chapter examines the challenges of current release processes in the automobile industry. It identifies, that the efficient and legally secured handling of changes is a crucial success factor. Therefore, state-of-the-art concepts for change propagation analysis and retest selection techniques are discussed and evaluated. It is determined that there exists no approach that is applicable for large systems at vehicle level that allows a reliable selection of all tests which are necessary to analyze the impact of the change. To address this problem two generic approaches are introduced and compared. Inclusion-based test selection concepts identify tests from the set of not executed tests and transform them into executed tests, whereas exclusion-based test selection techniques eliminate test cases from the set of executed tests. Both methods can achieve high inclusiveness and low error rates if their selection quality is ideal. The costs of the approaches depend on the weighting of misclassification costs. Therefore, no preferable approach can be determined in general. Thus, a further detailing is necessary. A concept for an exclusion-based test selection technique is presented, which eliminates test cases by the use of three exclusion arguments: result neutrality, simple result equality and multiple result equality. It offers the opportunity to reduce release test effort without drawbacks in test inclusiveness.

The upcoming development steps will be to further concretize the mentioned exclusion arguments. Actual research questions are:

- How can be formally proved that a change has no impact on a test case?
- How can the results of test cases be compared?
- How can the software be designed to support the discussed concept?
- How can the test suite be structured to facilitate the described approach?

Further on, a concept for an inclusion-based test selection technique has to be developed. This will enable a detailed analysis of the achievable selection qualities and costs of the different approaches. It may be possible that in some cases the selection effort for a distinct differentiation between necessary and obsolete tests is higher than the execution of the tests. In these situations it is less expensive to carry out the test rather than to perform the selection technique. Therefore, it may be useful to introduce further categories.

With these results, it will be possible to determine, if a test selection technique exists, that overcomes the problems of the state-of-the-art methods.

References

Aigner, J.: Zur zuverlässigen Beurteilung von Fahrzeugen. Automobiltechnische Zeitschrift (ATZ). **84**(9), 447–450 (1982)

Albers, A.: Vorwort zu Band 47 der Forschungsberichte des Instituts für Produktentwicklung, Karlsruhe (2010)

AUTOSAR (AUTomotive Open System ARchitecture): http://www.autosar.org (2014)

Borgeest, K.: Elektronik in der Fahrzeugtechnik. Vieweg+Teubner Verlag, Wiesbaden (2008)

Briand, L.C., Labiche, Y., He, S.: Automating regression test selection based on UML designs. ACM Trans. Softw. Eng. Methodol. **51**, 16–30 (2009)

Broy, M.: Challenges in automotive software engineering. In: Proceedings of the International Conference on Software Engineering (ICSE), Shanghai (2006)

Broy, M., Reichart, G., Rothhardt, L.: Architekturen softwarebasierter Funktionen im Fahrzeug: von den Anforderungen zur Umsetzung. Informatik-Spektrum. **34**(1), 42–59 (2011)

Burgdorf, F.: Eine kunden- und lebenszyklusorientierte Produktfamilienabsicherung in der Automobilindustrie. Dissertation, Karlsruher Institut für Technologie. KIT Scientific Publishing, Karlsruhe (2010)

Caliebe, P., Herpel, T., German, R.: Dependency-based test case selection and prioritization in embedded systems. In: 5th International Conference on Software Testing, Verification and Validation (ICST), Montreal (2012)

Chittimalli, P.K., Harrold, M.J.: Regression test selection on system requirements. In: India Software Engineering Conference (2008)

Clarkson, P.J., Simons, C., Eckert, C.: Predicting change propagation in complex design. In: Proceedings of ASME Design Engineering Technical Conferences and Computers and Information in Engineering Conference, DETC, Pittsburgh (2001)

Cohen, T., Navathe, S.B., Fulton, R.E.: C-FAR, change favorable representation. Comput. Aided Des. **32**, 321–328 (2000)

Drummond, C., Holte, R.C.: Explicitly representing expected cost: an alternative to roc representation. In: Proceedings of the 6th International Conference on Knowledge Discovery and Data Mining, pp. 155–164 (2000)

DSM Web: Design Structure Matrix. http://www.dsmweb.org/en/understand-dsm/tutorials-overview/descripton-design-structre.html (2014)

Düser, T.: X-in-the-Loop – ein durchgängiges Validierungsframework für die Fahrzeugentwicklung am Beispiel von Antriebsstrangfunktionen und Fahrerassistenzsystemen. Karlsruher Institut für Technologie, Forschungsberichte des IPEK – Institut für Produktentwicklung (2010)

Eckert, C., Clarkson, P.J., Zanker, W.: Change and customization in complex engineering domains. Res. Eng. Des. **15**, 1–21 (2004)

Eppinger, S.D., Whitney, D.E., Smith, R.P., Gebala, D.A.: A model-based method for organizing tasks in product development. Res. Eng. Des. 1–13 (1994)

Economic Commission for Europe of the United Nations: ECE R13 H: Uniform provisions concerning the approval of passenger cars with regard to braking (2015)

Flanagan, T.L., Eckert, C., Smith, J., Eger, T., Clarkson, P.J.: A functional analysis of change propagation. In: Proceedings of the 14th International Conference on Engineering Design (ICED'03), Stockholm (2003)

Fürst, S.: Challenges in the design of automotive software. In: Proceedings of the Conference on Design, Automation and Test in Europe (DATE), Dresden (2010)

Gallagher, K., Hall, T., Black, S.: Reducing regression test size by exclusion. International Conference on Software Maintenance, IEEE, pp. 154–163 (2007)

Gorthi, R.P., Pasala, A., Chanduka, K.K.P., Leong, B.: Specification-based approach to select regression test suite to validate changed software. In: 15th Asia-Pacific Software Engineering Conference (2008)

Gustavsson, H.: Architecting automotive product lines. In: Bosch, J., Lee, J. (eds.) SPLC 2010, LNCS 6287, pp. 92–105. Springer Verlag, Berlin (2010)

Holzmann, H., Hahn, K.M., Webb, J., Mies, O.: Simulationsbasierte ESP-Homologation für Pkw. Automobiltechnische Zeitschrift ATZ, Ausgabe 09/2012, pp. 698–702 (2012)

HTWK Leipzig: V-Modell. http://www.imn.htwk-leipzig.de/~weicker/upload//Main/V-Modell.png (2014)

International Automotive Task Force: IATF 16949: Quality management system requirements for automotive production and relevant service parts organizations (2016)

International Organization for Standardization: ISO 26262: Road vehicles – Functional Safety (2011)

Jarratt, T., Eckert, C.M., Clarkson, P.J.: Development of a product model to support engineering change management. In: Proceedings of the TCME 2004, Lausanne (2004)

Kaner, C.: Improving the maintainability of automated test suites. In: Proceedings of Quality Week (1997)

Keller, R., Eger, T., Eckert, C.M., Clarkson, P.J.: Visualizing change propagation. International Conference on Engineering Design (ICED'05), Melbourne (2005)

Khan, S., Rehman, O., Malik, S.: The impact of test case reduction and prioritization on software testing effectiveness. In: International Conference on Emerging Technologies, ICET, pp. 416–421 (2009)

Kim, J., Porter, A., Rothermel, G.: An empirical study of regression test application frequency. In: Software Testing, Verification and Reliability, vol. 15, pp. 257–279 (2005)

Kohn, W.: Statistik – Datenanalyse und Wahrscheinlichkeitsrechnung. Springer, Berlin (2005)

Kvasnicka, P., Prokop, G., Dörle, M., Rettinger, A., Stahl, H.: Durchgängige Simulationsumgebung zur Entwicklung und Absicherung von fahrdynamischen Regelsystemen. In: Berechnung und Simulation im Fahrzeugbau, Würzburg, VDI-Berichte 1976, pp. 387–404, VDI-Verlag, Düsseldorf (2006)

Langermann, R.: Beitrag zur durchgängigen Simulationsunterstützung im Entwicklungsprozess von Flugzeugsystemen. Dissertation, TU Braunschweig (2008)

Lehnert, S.: A review of software change impact analysis. TU Ilmenau, Report ilm1-2011200618 (2011)

Leung, H., White, L.: A study of integration testing and software regression at the integration level. In: Microsoft Research (Eds.) – TechReport MSR-TR-2005-94 (2005)

Mao, Y., Wiessalla, J., Meier, J., Risse, W., Mathot, G., Blum, M.: CAE supported ESC development/release process. In: Proceedings of the FISITA 2012 World Automotive Congress (2012)

Muccini, H., Dias, M., Richardson, D.J.: Software architecture-based regression testing. J. Syst. Softw. **79**(10), 1379–1396 (2006)

National Highway Traffic Safety Administration (NHTSA): Federal Motor Vehicle Safety Standard (FMVSS) No. 126: Electronic Stability Control Systems (2007)

Nörenberg, R.: Effizienter Regressionstest von E/E-Systemen nach ISO 26262. Dissertation, Karlsruher Institut für Technologie (KIT). KIT Scientific Publishing, Karlsruhe (2012)

Object Mangament Group (OMG): Unified Modeling Language (UML). http://www.uml.org/ (2014)

Raffaeli, R., Germani, M., Graziosi, S., Mandorli, F.: Development of a multilayer change propagation tool for modular products. In: International Conference on Engineering Design, ICED, Paris (2007)

Rausch, A., Broy, M.: Die V-Modell XT Grundlagen. In: Höhn, R., Höppner, S. (eds.) Das V-Modell XT, pp. 1–27. Springer, Berlin (2008)

Reif, K.: Automobilelektronik. ATZ/MTZ-Fachbuch. Vieweg Verlag, Wiesbaden (2007)

Reuter, A.: Produkthaftung in Deutschland. In: Werdich, M. (ed.) FMEA – Einführung und Moderation. Vieweg+Teubner Verlag, Wiesbaden (2011)

Rothermel, G., Harrold, M.J.: Analyzing regression test selection techniques. IEEE Trans. Softw. Eng. **22**(8), 529–551 (1996)

Rothermel, G., Harrold, M.: Empirical studies of a safe regression test selection technique. IEEE Trans. Softw. Eng. **24**, 401–419 (1998)

Rothermel, G., Harrold, M.J., von Ronne, J., Hong, C.: Empirical studies of test-suite reduction. J. Softw. Testing Verification Reliability. **12**(4), 219–249 (2002)

Schäuffele, J., Zurawka, T.: Automotive Software Engineering. Vieweg Verlag, Wiesbaden (2003)

Singer, C.: Entwicklung von Testauswahlmethoden für die Absicherung von Änderungen auf Gesamtfahrzeugebene. VDI Fortschritt-Berichte, Reihe 12, Nr. 798. VDI-Verlag, Düsseldorf (2016)

Steward, D.: The design structure system: a method for managing the design of complex systems. IEEE Trans. Eng. Manag. **EM-28**(3), 71–74 (1981)

Sundmark, D., Petersen, K., Larsson, S.: An exploratory case study of testing in an automotive electrical system release process. In: Proceedings of the 6th IEEE International Symposium on Industrial Embedded Systems (SIES), Västerås (2011)

Tan, P.-N.: Receiver operating characteristic. In: Liu, L., Özsu, M.T. (eds.) Encyclopedia of Database Systems, pp. 2349–2352. Springer, New York, NY (2009)
Test Glossary.: http://testingstandards.co.uk/glossary.htm (2014)
Ungermann, J.: Zuverlässigkeitsnachweis und Zuverlässigkeitsentwicklung in der Gesamtfahrzeugerprobung. Dissertation, ETH Zürich (2009)
VDA QMC: Automotive SPICE Process Assessment/Reference Model v3.0 (2015)
Vokolos, F., Frankl, P.: Pythia: a regression test selection tool based on text differencing. In: Proceedings of the International Conference on Reliability, Quality, and Safety of Software Intensive Systems (1997)
Yazdanshenas, A.R., Moonen, L.: Fine-grained change impact analysis for component-based product families. In: International Conference on Software Maintenance (ICSM), IEEE (2012)
Zhao, O., Yan, H., Xiang, L., Xu, B.: Change impact analysis to support architectural evolution. J. Softw. Maintenance Res. Pract. – Special Issue: Separation of Concerns for Software Evolution 317–333 (2002)
Zheng, J., Robinson, B., William, L., Smiley, K.: An initial study of a lightweight process for change identification and regression test selection when source code is not available. In: 16th International Symposium on Software Reliability Engineering (2005)

Part II
Requirement Analysis and Systems Architectures

Chapter 3
Increasing Energy-Efficient Driving Using Uncertain Online Data of Local Traffic Management Centers

Per Lewerenz and Günther Prokop

Abstract The main goals of today's research and development are leading to different systems and topics for more energy-efficient technologies in powertrains and intelligent driver assistance systems. The funded project "Energieeffizientes Fahren 2014" (EFA 2014/2) aims for increasing the electric vehicles' operation range. In order to reach this goal an approach has been chosen which includes infrastructure data using Vehicle-to-Infrastructure (V2I) communication technologies. Particularly traffic actuated traffic lights are being utilized since this is state of the art to optimize traffic flow. Based on the interaction between vehicle and infrastructure the driver will be able to achieve an energy-efficient manner of driving through additional information and integrated board aggregation. This approach has been successfully tested in Dresden.

Keywords Traffic management • Communication • Car2X • HMI • Energy efficient • Microscopic traffic simulation • Dresden

3.1 Motivation

Individual vehicle traffic is increasing constantly and with this emission, which leads to higher environmental pollution. In order to reduce increasing emissions due to road traffic, electric vehicles are being utilized. The disadvantage of these vehicles is the relatively short range of operation. The Research and Development (R&D) project EFA 2014/2, funded by the Federal Ministry of Education and Research aims to increase the range by adapting energy-efficient driving behaviors, e.g. precise speed recommendations while approaching a traffic light. Therefore an advanced driver assistance application has been developed. It comprises the interaction between a vehicle and its surroundings—especially

P. Lewerenz (✉) • G. Prokop
Institut für Automobiltechnik Dresden, Lehrstuhl für Kraftfahrzeugtechnik, George-Bähr-Straße 1c, 01062 Dresden, Germany
e-mail: per.lewerenz@tu-dresden.de

H. Winner et al. (eds.), *Automotive Systems Engineering II*,
DOI 10.1007/978-3-319-61607-0_3

traffic lights—based on Vehicle-to-Infrastructure (V2I) communication since speed recommendations are only possible by knowing upcoming signal states of the traffic lights ahead.

3.2 Online Infrastructure Data Sources

Traffic management actions require a lot of information that can be obtained e.g. from induction loops, cameras and floating car data (FCD). Utilizing this detected data, infrastructure components as traffic lights or variable message signs can be influenced and adapted in order to affect the current traffic flow. Since energy consumption is dependent on traffic flow, which is mainly controlled by traffic lights in urban areas, it is necessary to understand how traffic light controls work. A new approach is being established which aims to change the behavior of a single vehicle instead of the traffic flow. Therefore the ascertainment of an individual energy-efficient driving behavior for every single vehicle is being pursued depending on traffic signal states. Signal times and delay times, as the most important characteristics, are being presented in the following chapters.

3.2.1 Prediction of Signal States

In order to determine an energy-efficient driving behavior while approaching a traffic light, it is essential to know the signal state at the time the vehicle reaches the stop line. This implicates the necessity of predicting future signal times. Since most of the existing traffic lights are traffic actuated, signal times are not fixed and the use of probability based methods is required (Krumnow 2012). Traffic actuated traffic light controls contain different signal programs depending on traffic demand which varies from morning over midday to evening hours (Fig. 3.1).

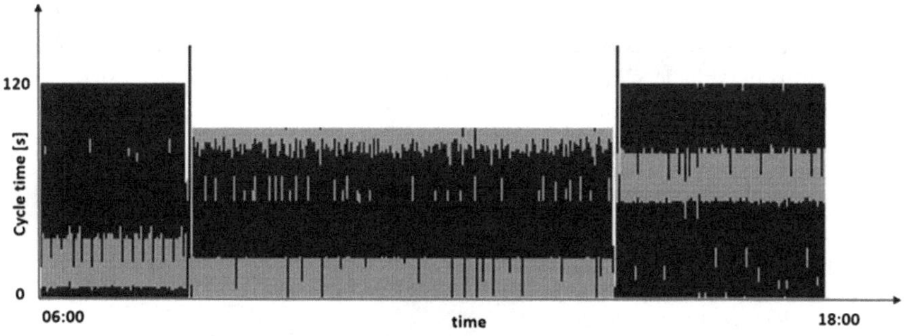

Fig. 3.1 Signal states of a traffic actuated signal program

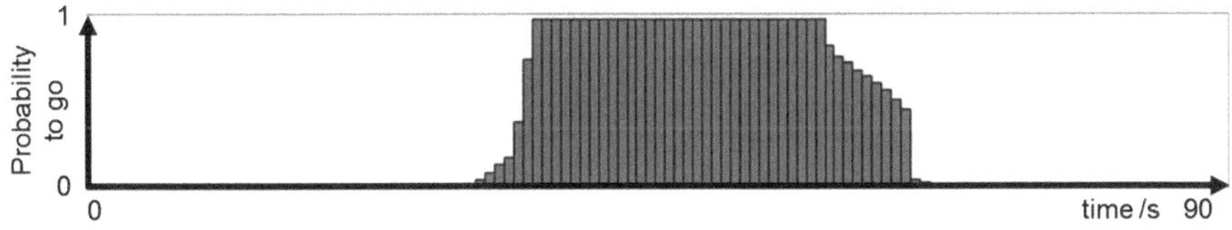

Fig. 3.2 Example of "probability-to-go" values

Within these signal programs it is possible to request additional phases for public transport or little frequented directions causing shifted signal times.

To ascertain the predicted probabilities to go, signal states are being simplified to only two states and coded in binary vectors. Therefore, "0" relates to the states red, amber and red-amber, and "1" refers only to the state green (Krumnow 2014). This vector composed over a time period of several minutes is being transmitted to the vehicle where energy-efficient driving behavior is being computed (Fig. 3.2).

3.2.2 Delays Due to Traffic Light Signals

Another important fact is the knowledge of congestion in front of traffic light controlled intersections. There are only limited possibilities measuring the queue length, e.g. with cameras, so it is more common to estimate the values of queue lengths. The used estimation procedure depends on the disposability of data sources. Statistic and heuristic methods are being applied in case only historic or aggregated data is available. If data of fixed detectors e.g. induction loops are available, approaches like the method of Mück (2002) can be utilized.

If mobile sensors like floating car data are the basis for the prediction, the method of Neumann (2011) can be applied. Within this project static sensors are being used in order to estimate delay time in seconds and queue length in meters. Both values are being updated every second. In addition to signal times, also congestion affects the traffic light control so that the knowledge of queue length and delay time is essential.

3.3 Communication Chain and Car Positioning

This part describes the fundamental and technical challenges within the communication chain from traffic lights into the vehicle. In Particular, this includes the latency of required information between the traffic lights and the car. Furthermore the direct conjunction of the vehicle position and the suitable information in urban scenarios is also a challenge.

These aspects can be well explained by the following example of a driving situation at a multitrack intersection in an urban scenario. The vehicle positioning solution based on GNSS is too vague for accurate localization caused by strong multipath urban environment. Due to this, the positioning solution does not allow a relation to the real driven track. In this case, the car computer does not discern which traffic light information (e.g. red or green remaining time) should be displayed for the driver. That said, the driver gets all possible traffic light information and needs to choose the right one. To avoid this, a track selection is needed. Furthermore, the displayed traffic light has to be the right one. In order to do so, the whole technical system has to handle the delayed information over the complete communication chain between traffic light and car.

For communication different types of systems (e.g. GSM/UMTS) and protocols (e.g. TPEG TSI—Transport Protocol Experts Group Traffic Service Information) are used. A complete reproduction of the communication chain in a laboratory allows simulation and measurements of the latency between different parts of the chain. Therefore, testing of different latency time measurements and interference scenarios and their solutions for communication becomes possible.

Two approaches, Kalman- and Particle-Filters, have been examined for car positioning. These two filters are state of the art (Bar-Shalom et al. 2001) and were combined with an enhanced digital map and preprocessed video data (Gosda et al. 2013).

The video data contains information about distances to the left and the right lane marks. Furthermore, the enhanced digital map provides all lanes for the test scenario in Dresden with additional information like stop markings. This information is deposited as an XML scheme and can be easily virtualized, combined and overlain with other maps. Figure 3.3 gives an overview of the used car sensors and sensor data. The data is provided via CAN bus to the car computer and is synchronically sampled for the filters. The idea is to run two filters in parallel for using

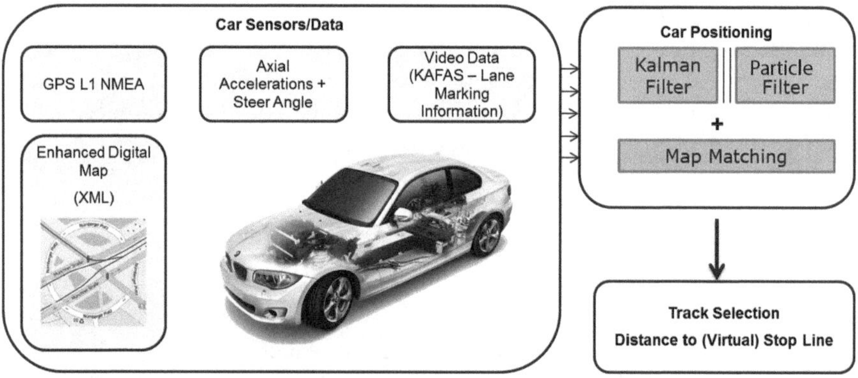

Fig. 3.3 Overview of car sensors/data (KAFAS—Camera Assisted Driver Assistance System, NMEA—National Marine Electronics Association) and the positioning algorithms for track selection and distances to (virtual) stop lines

estimated solutions for track selection. Combined with additional video data, different hypotheses for lane selection will be examined based on the enhanced digital map. In order to assess the performance of these methods, synthetic data for filter calibration is used at first. After that, real data from the test scenario environment is used for validation. Due to this, a precision ($\leq$0.5 m for $\pm 2\sigma$) can be achieved for car positioning in case of the test environment in Dresden. The value 2σ means the probability of the positioning values is better or equal than 0.5 m in 95.4% of cases. In 4.6% of the cases the probability of the positioning values is worse than 0.5 m.

3.4 Real Traffic Investigation

This chapter shows the way how traffic information presented above can be displayed in a real vehicle to the driver, leading to optimal driving behavior.

3.4.1 Experimental Vehicle

To validate the developed measures in real traffic, an experimental vehicle is used. This is a full electric car manufactured by BMW called ActiveE. A picture of the vehicle can be seen in Fig. 3.4. It is a converted BMW 1 Series Coupe (E82e). The electric motor drives the rear axle with a maximum power of 125 kW. The vehicle has a lithium-ion-battery with a capacity of 32 kWh which lasts for driving ranges up to 160 km.

The experimental vehicle is modeled in MATLAB/SIMULINK, to have the opportunity to simulate the energy consumption of the vehicle at different traffic conditions. Additional information on the model can be found in Schubert et al. (2014).

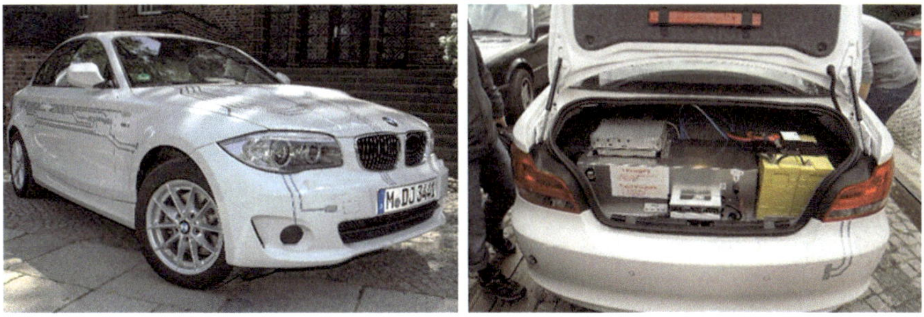

Fig. 3.4 Experimental vehicle with in-vehicle measurement system

3.4.2 Human Interaction

By merging vehicle internal and infrastructure information in future driver assistance functions, the amount of information for drivers will increase significantly, as the present research project shows.

It is important to make sure that the information does not demand too much attention from the driver (Winner et al. 2012). Furthermore, it should be noted that due to the source, the information will not always be reliable. Therefore, the forms of representation must be chosen so that the driver assesses the information as helpful even if they are uncertain. This means that the form of representation has a significant impact on the overall acceptance of the system. The approaches are shown to communicate uncertain information for efficient driving in traffic light approach situations to the driver. As information channels a visual display in the dash panel and haptic feedback via an active accelerator pedal are used. The installed dash panel of the used test vehicle can be programmed to flexibly examine different forms of representation. Near-series illustrations are presented to the driver and the impact on driver behavior through prototypical auxiliary displays can be avoided. With help of the active accelerator pedal, the counter force, which the driver has to apply to the operation of the pedal, can be varied dynamically. On the one hand, a direct influence of the driver can be initiated; on the other hand, the driver's attention can be stimulated by vibrating the accelerator pedal. For the optical information two areas were defined in the display. One area is used for the recommendation, and the other one for information (Fig. 3.5).

In the area of recommendation a range of traveling speed is recommended. With the recommended speed, the next traffic light is reached at a green phase without vehicle standstill. The quality of information is taken into account that a distinction is made between the core times (probability-to-go > 90%) and a region of high probability (90% > probability-to-go > 70%). The representation in the information area depends on the current speed relative to the recommended speed. Therefore, the consideration of information quality in the information area is set automatically. If the speed of the vehicle is below a certain level (10 km/h), the remaining phase duration is displayed. The assignment and the information displayed are shown in Fig. 3.6.

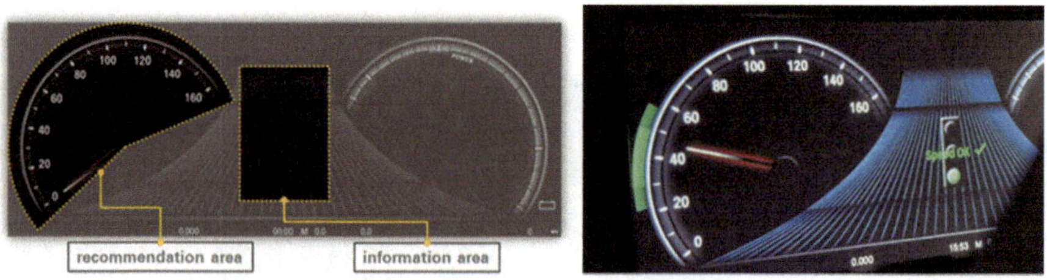

Fig. 3.5 Programmable digital instrument panel

Matching the information of the different areas, the active accelerator pedal is triggered. In cases (1) and (5) the driver is informed by a slight vibration of the pedal that he should optimize its longitudinal dynamic behavior. In cases (2) and (4) the driver is caused by the increase or decrease of the pedal counter force to change the speed to the optimal range.

3.4.3 *Validate the Benefit of Driver Assistance in Simulated Traffic Scenarios*

To get reliable results while validating the effect of driver assistance systems in real traffic a huge amount of test kilometers is necessary. Simulation is a very powerful way to analyze these systems.

Current researches in Bley et al. (2011), Schubert (2010) and Schuricht et al. (2011) in the context of traffic light assistance systems (TLAS) and predictive cruise control systems (Asadi and Vahidi 2010) show the high potential of driver assistance systems to realize an energy efficient driving behavior. All of these systems use the information of traffic lights to calculate an optimal velocity to approach the intersection. Consumption reductions between 3% and 5% (Bley et al. 2011) and in some situations about 30% (Schuricht et al. 2011) in comparison to the uninformed driver are shown. For a general declaration it is essential to examine the influence of other road users, in particular the queue length, at the stop line of traffic lights. Currently the scenarios examined are only very simple traffic situations (e.g. one lane, static traffic light programs).

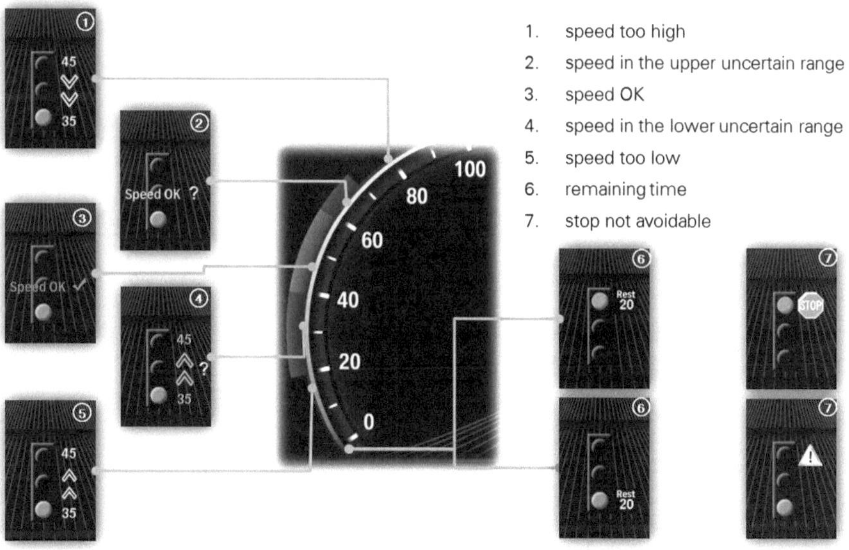

Fig. 3.6 Assignment of information and recommendation

To simulate more complex traffic scenarios a new simulation framework (Schubert et al. 2013) was developed. It is an interface between detailed nanoscopic vehicle simulation with MATLAB/Simulink and traffic flow simulation with SUMO (Simulation of Urban Mobility). With different parameters like speed limits, traffic light control, number of lanes, and density of traffic flow, the impact of different traffic situations on the energy consumption can be determined. For that purpose a vehicle model (see Sect. 3.5) is used to determine the energy consumption. Different traffic scenarios and their influence on the individual energy consumption are analyzed. The results will show whether these systems are useful and how beneficial it is.

3.5 First Results

In a first realistic use case the potential of the traffic light assistant TLAS is shown. A part of an urban route in Dresden with a constant traffic stream is modelled in the SUMO Simulation suite. On basis of Asadi and Vahidi (2010) a traffic light assistance system is implemented in MATLAB and simulated with the framework described in Sect. 3.4.3. Different parameters e.g. the time of occurrence of the analyzed vehicle are varied.

Besides the reduction of vehicle stops and trip time the energy consumption of a vehicle is important. It is calculated with a model (Schubert et al. 2014) of the described vehicle (see Sect. 3.4.1). Figure 3.7 shows the trajectories of a vehicle for different times of occurrence in the provided road network. Evidently, some of the vehicle stops can be prevented.

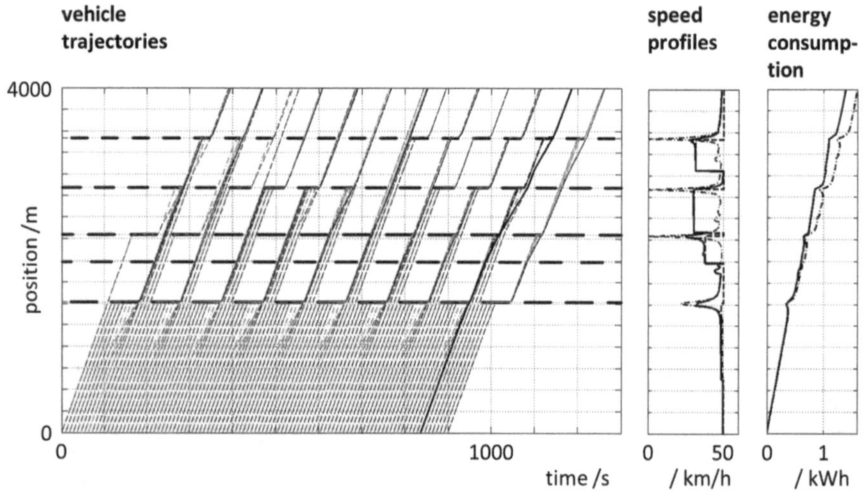

Fig. 3.7 Simulation results: vehicle trajectories, detailed speed profile and energy consumption

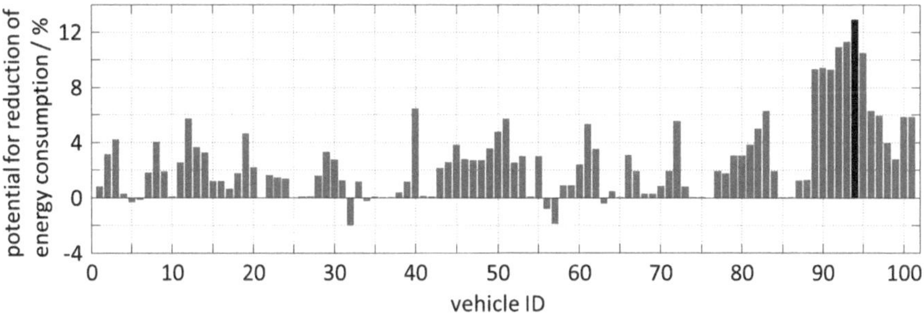

Fig. 3.8 Simulation results: energy consumption of simulated traffic situations. Vehicle 94 is highlighted with the best potential

The consumption for each of these vehicles can be seen in Fig. 3.8. The highest potential for energy consumption is provided for vehicle number 94. Although there are obviously some disadvantageous situations that could cause higher energy consumption (see vehicles 32, 56 and 57 in Fig. 3.8).

These are caused by the effect that in some cases a vehicle with TLAS could pass the next traffic light but this leads to a disadvantageous situation at the next traffic light, so the TLAS cannot avoid the vehicle stop. This demonstrates the need of further research on that topic.

3.6 Conclusions

The R&D project EFA 2014/2 is pursuing the approach to obtain infrastructure information of the whole road network, e.g. traffic light information. This information is being broadcasted using the widespread and available mobile communications network. This involves assets and drawbacks, e.g. concerning the availability and higher latency times compared to usual directional connections between car and traffic light. In urban areas traffic actuated traffic lights are being utilized now and in the future even more. These traffic light controls are being influenced by traffic flows as well as single vehicles so that it is possible to react dynamically to varying traffic demands. Due to this variation, only a prediction approach is capable. In doing so, it is necessary to create an approach that is able to deal with probability-based values.

Results of various simulations show an energy-saving potential of about 10% by using this developed approach. To investigate the approach in real traffic, a test vehicle was equipped with the system. Therefore, a special display has been developed that is able to deal with probability-based values. In order to achieve the desired driving strategy the driver is being supported actively by the system. On the one hand, the driver gets driving instructions via a display; on the other hand, an onboard unit pays attention to an efficient realization of these instructions. Finally an energy-efficient approach has been created to increase the range of electric vehicles which is transferable to other cities.

In future, needed data streams can be distributed over different communication systems. Therefore, non-time-critical information (e.g. map data) can be distributed by broadcast. For time-critical information (e.g. short-term prediction data) it seems to be suitable to use communication systems with low latency. The state of the art is to optimize driving strategies while approaching single traffic lights. Considering all traffic lights on a route may lead to even higher saving potentials. Also the developed energy-efficient driving strategy needs to be adapted to these new conditions. Nevertheless, it is conceivable that a wide transmission of signal data can lead to a global optimum regarding to the driving strategy for the whole route.

An important point is that driver assistance systems will capture the infrastructure more effectively in future, but also the infrastructure systems will react smarter to the traffic.

Acknowledgments This article is based on the results of the project EFA 201/2, which was funded by the Federal Ministry of Education and Research. The author would like to thank the colleagues Mario Krumnow, Robert Richter and Torsten Schubert, who have contributed to the project.

References

Asadi, B., Vahidi, A.: Predictive cruise control: utilizing upcoming traffic signal information for improved fuel economy and reducing trip time. IEEE Trans. Control Syst. Technol. **19**, 707–714 (2010)

Bar-Shalom, Y., Li, X., Kirubarajan, T.: Estimation with Applications to Tracking and Navigation: Theory Algorithms and Software. Wiley, New York (2001)

Bley, O., Kutzner, R., Friedrich, B.: Kooperative Optimierung von Lichtsignalsteuerung und Fahrzeugführung. In: AAET 2011 Automatisierungs-, Assistenzsysteme und eingebettete Systeme für Transportmittel (2011)

Gosda, U., Weber, R., Michler, O., Zeisberg, S., Mademann, E.: Target tracking in wireless sensor networks by data fusion with video-based object detection. In: 10th Workshop on Positioning Navigation and Communication (WPNC), pp. 20–21 (2013)

Krumnow, M.: Schaltzeitprognose verkehrsadaptiver Lichtsignalanlagen im Rahmen des Projektes EFA 2014/2, 8. VIMOS Tagung, Dresden (2012)

Krumnow, M.: Schaltzeitprognose verkehrsabhängiger Lichtsignalanlagen im Rahmen des forschungsprojektes EFA 2014/2, Tagungsbericht Heureka 2014. FGSV Verlag, Köln (2014)

Mück, J.: Schätzverfahren für den Verkehrszustand an Lichtsignalanlagen unter Verwendung halteliniennaher Detektoren, Tagungsbericht Heureka 2002. FGSV Verlag, Köln (2002)

Neumann, T.: Rückstaulängenschätzung an Lichtsignalanlagen mit Floating-Car-Daten. Berichte aus dem DLR-Institut für Verkehrssystemtechnik, Berlin (2011)

Schubert, T.: Entwurf und Evaluierung einer prädiktiven Fahrstrategie auf Basis von Ampel-Fahrzeug-Kommunikationsdaten. Diplomarbeit, Technische Universität Dresden (2010)

Schubert, T., Krumnow, M., Bäker, B., Krimmling, J.: Using nanoscopic simulations to validate the benefit of advanced driver assistance systems in complex traffic scenarios. In: Proceedings of the 3rd International Conference on Models and Technologies for Intelligent Transportation Systems 2013, Verkehrstelematik, vol. 3. TUDpress, Dresden (2013). ISBN: 978-3-944331-34-8

Schubert, T., Uebel, S., Krumnow, M., Bäker, B., Krimmling, J.: Analyse eines LSA-Assistenzsystems mittels nanoskopischer Simulation in komplexen Verkehrsszenarien. AAET – Automatisierungssysteme, Assistenzsysteme und eingebettete Systeme für Transportmittel, Braunschweig (2014)

Schuricht, P., Michler, O., Bäker, B.: Efficiency-increasing driver assistance at signalized intersections using predictive traffic state estimation. In: 14th International IEEE Conference on Intelligent Transportation Systems (ITSC), pp. 347–352 (2011). doi:10.1109/ITSC.2011.6083111

Winner, H., Hakuli, S., Wolf, G.: Handbuch Fahrerassistenzsysteme. Vieweg+Teubner Verlag, Wiesbaden (2012)

Chapter 4
Modelling Logical Architecture of Mechatronic Systems and Its Quality Control

Alarico Campetelli and Manfred Broy

Abstract In this work an integrated method for the development of mechatronic systems is presented for capturing information from requirements to code generation level, with derived and intermediated abstractions in a logical view. Our modelling theory, based on FOCUS, is a model-based engineering method for the development of reactive software systems. It supports the specific needs of the automation and automotive domains, and provides a model-based logical representation of the system together with a user-friendly integration of automatic verification. The scope of this work is to present our model-based development methodology for mechatronic systems, which provides an integrated way to define the respective engineering process models and formalisms, system requirements and architectures, to specify the behaviour of the system. Therefore, novel complementary analysis techniques can be applied, allowing the verification of properties, the validation of system design and derived model-based implementations. Moreover, a wider support for discipline neutral models reduces errors during integration of artefacts from individual disciplines.

Keywords Interactive systems • I/O-machines • Mechatronic systems • Model-based development • Modelling • Formal verification • Hybrid systems • Sampling

4.1 Introduction

Nowadays, safety-critical embedded systems are in use in vehicles, machines, aircrafts, and medical devices. At the same time, the role of the software in such system is rapidly increasing, determining the needs for integrated and multidisciplinary development processes. The following characteristics of mechatronic systems determine complex design challenges: big product portfolio, strong dependency in the

---

A. Campetelli
München, Germany

M. Broy (✉)
Institut für Informatik, Technische Universität München, Boltzmannstr. 3, 85748 Garching, Germany
e-mail: broy@in.tum.de

H. Winner et al. (eds.), *Automotive Systems Engineering II*,
DOI 10.1007/978-3-319-61607-0_4

development process between the involved disciplines and in the products between the internal components, and short product life cycles. These characteristics make development phases particularly difficult. For instance, the evolution of system models for new features or corrections are especially error-prone and induce high quality assurance costs. Moreover, design decisions during the development are difficult, because often not all impacts are known at decision time.

A fundamental challenge is to bring together principles, models, and development processes from mechanical engineering, E/E engineering and software engineering. Model-based development approaches are getting more widely used in software engineering. An important challenge is to introduce the principles and methodologies that are consolidated in the software domain to the design methods of other disciplines involved in the design and realization of mechatronic systems. Our long-term vision is an integrated modelling approach, which consistently combines sub-models from different disciplines in a common modelling framework. A description of the system that is fully discipline neutral is not possible at each stage of the system life cycle. However, a separation between disciplines only later in the development guarantees a less difficult successive integration phase. Therefore, covering as much as possible system artefacts through models supported by software tools permits to take advantage of the model-based principles, as functional-oriented design, simulation, verification in early design phases, and a more agile product versioning and variation.

Innovation and increment of functionalities are crucial challenges for mechatronic systems also determining an increase in the overall costs and system complexity. The introduction of model-based principles for a more optimised development processes can help mitigating these problems. System engineering needs to be focused on requirement engineering, architecture design and integration in a structured and automated way with a seamless use of system models this way guaranteeing comprehensive quality assurance.

In this work, we present a model-based development methodology for mechatronic systems. For the classification of the system artefacts, we refer to the so-called SPES matrix, which has been developed in the research projects SPES 2020 and SPES XT (Pohl et al. 2012)[1] see Fig. 4.1 as reference. The columns of the matrix are the artefact viewports. Viewports collect artefacts specific for requirements, functional, logical and technical level and the rows of the matrix their levels of granularity.

FOCUS (Broy et al. 1992) is a formal modelling theory for the formal specification of distributed, discrete-event systems. It forms the theoretical foundation for the models of the functional and logical viewpoints of our methodology. This formalism defines architecture of systems through a hierarchical and interconnected net of components, with i/o typed interfaces. The internal behaviour of each component can be implemented using different formalisms, e.g. functional specifications or finite state machines. A formal modelling theory, as FOCUS, provides important concepts for development such as a strict subdivision of different

[1]http://spes2020.informatik.tu-muenchen.de/

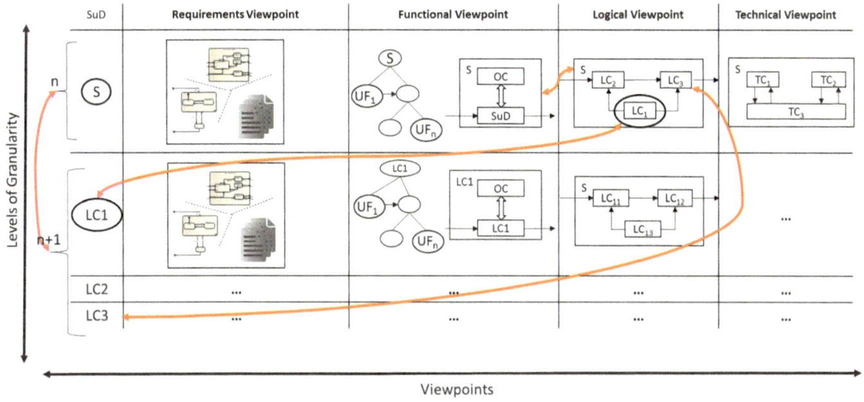

Fig. 4.1 Two-dimensional abstraction SPES matrix (Pohl et al. 2012): levels of granularity and artefacts viewpoints. *SuD* system under development, *S* proper subset of the system, *OC* operational context, *UF* user function, *LC* logical component, *TC* technical component

conceptional levels of the system, views and different levels of granularity of the system under development. Its comprehensive use permits the construction of more reliable systems, even though system complexity is an issue, which is expected to even grow in the future. Moreover, a suitable modelling theory for mechatronic systems helps in their development, maintenance, simulation, and verification. Correct behaviour of the modelled systems is fundamental. It can be guaranteed through analysis techniques, mostly in form of formal verification. The use of formal methods for the system definition allows for formal verification techniques, which offer exhaustive and automatic checks. AutoFOCUS 3[2] is a research tool for modelling. Its semantics is founded on the FOCUS theory.

In the automotive and automation domains, and more general for mechatronic systems, the combination of embedded systems with physical components requires a suitable design that unites continuous and discrete behaviour. The most important challenge lies in modelling the interaction between system and the environment with its physical constraints. In fact, software components operate in discrete program steps, while the physical components function over continuous time intervals following physical constraints. In software engineering, systems that combine discrete/continuous time and data are called hybrid systems. Software-intensive mechatronic systems modelled in hybrid systems are complex: These systems consist of a high number of modules or programs, where the software part is in the magnitude of several millions lines of code. System failures may lead to a considerable loss of money due to warranty costs or even—in the worst case—endanger human lives. This motivates the need for well-defined formal modelling theories, languages, and tools, which help to improve system quality.

[2]http://autofocus.in.tum.de

A widely used paradigm for modelling hybrid systems is the hybrid automaton (Henzinger 1996). Originally, the FOCUS theory has been introduced with a discrete time model of computation. In this work, we present recent extensions to a continuous time model of computation (Campetelli 2013; Broy 2012). We introduce i/o hybrid state machines to FOCUS, inspired by the hybrid automaton formalism, and address aspects related to their simulation and formal verification.

Numerous model-based approaches aiming to overcome the criticalities by the development of mechatronic systems. The MechatronicUML language supports the specification of component-based software in mechatronic systems and is introduced by Becker et al. (2014). It is based on the concepts of UML supporting structural and behavioural aspects. Kernschmidt and Vogel-Heuser (2013) present an interdisciplinary modelling approach called SysML4Mechatronics. In this approach are analysed and modelled evolutions and changes during the development mechatronic systems, within one engineering discipline as well as between different disciplines. A similar approach called 3 + 1 SysML integrates in a common view-model modelling artefacts relative to mechanical, electronic and software parts of mechatronic systems (Thramboulidis 2010). Habib (2007) states the need to introduce founded theories and tools in the process of development of mechatronic systems. The author argues that mechatronic engineering should become an autonomous discipline, in which data and models of single disciplines should be integrated. In order to analyse current development process, challenges and future trends of mechatronic systems, Schäfer and Wehrheim (2007) survey the process of development of a rail system. The authors identify the importance of an integrated approach between the involved engineering disciplines to deal with adaptively, self-coordination and self-organisation of mechatronic systems. These requirements determine in particular for software engineering important challenges in modelling, code generation and analysis. El-khoury et al. (2005) address the integration of different engineering disciplines proposing an architecture that supports integration and data exchange of models defined in different tools for different aspects of the system. Anacker et al. (2011) integrate mechatronics in the engineering process introducing a language for the specification of mechatronic systems and the support for reusable solution patterns.

The contribution is structured as follows: In the next section, we present our system design methodology. In the third section, we introduce the FOCUS design approach and in the fourth section, we explain the FOCUS modelling theory and its extension to a continuous time model of computation formally. Section 4.5 introduces the simulation and discretisation of FOCUS components with sampling techniques. The formal verification approach is presented in Sect. 4.6. Finally, in the last section we give summary and concluding remarks.

4.2 Methodology for the Design of Mechatronic Systems

The increasing complexity of software-intensive mechatronic systems due to more advanced functionalities and domain specific applications together with an intensive interaction between mechanical and E/E components with the physical environment. At the same time, the development process requires the collaboration of different engineering experts in an increasing short product life cycle, leading to involved integration phases between the discipline specific models. A seamless design methodology is required that provides suitable system models, processes, and methods to optimise the system engineering, together with exhaustive analysis techniques.

In research project SPES, sponsored by the German ministry of research, and its successor SPES XT, a large consortium of industrial and academic partners defined a method for model-based development of embedded systems, with the scope to develop model-based engineering methods that support verification, validation, handling of contractors, quality assurance, definition of system modes and systematic reuse.

The SPES development framework, also called the SPES matrix, provides modelling concepts to design different aspects of the system, e.g. for embedded systems its functional behaviour, its software architecture and its hardware architecture. The first dimension of the SPES matrix defines four viewpoints: requirements, functional, logical, and technical. The viewpoints compose the SPES methodology for the definition of system architecture. The order already suggests phases of the design process, where the artefacts of one viewpoint serve as input for the artefacts of the next viewpoint. However, the SPES framework does not prescribe a fixed development process: an evolution or change of a model in a viewpoint, it may require modifications to models in others viewpoints, in order to maintain the consistency between system representations. The second dimension of the matrix is the level of abstraction or granularity of the system under development (SuD in Fig. 4.1). In fact, the development of mechatronic systems usually requires different levels of granularity, to manage and to reduce the overall complexity. The SPES matrix is illustrated in Fig. 4.1.

The design of a system under development usually starts with its requirements, which can have several representations (e.g. textual or graphical) and levels of formality (e.g. formal or informal). An initial requirements management build the first draft of the functionalities and system boundaries. According to the SPES approach the requirements artefacts are collected in the first viewpoint and can be represented for instance by use case diagrams in terms of message sequence charts representing scenario descriptions. The functional viewpoint describes the system functions in a structured and hierarchical way looking at the system as a black box modelling the interface and behaviour of the system. The functional viewpoint can be directly derived from the system requirements. The artefacts, in this viewpoint, describe the system by a set of functions, which has to be realized by the system implementation. This way, it is possible to handle product lines and the evolution of the system functions, abstracting from implementation details. The next viewpoint is the logical viewpoint, which defines the system architecture from a logical point of view.

The artefacts in the logical viewpoint sketch a first structuration of the system into a hierarchy of interconnected units, called components, with defined i/o interfaces and behaviour. Functions are seamlessly traced with an association to the correspondent subsystems in the logical models that implement them. The behaviour of the system is observable in different ways, as for instance through observations of the signal/message flow between the components. Therefore, it is possible to systematically analyse the behaviour of the system and verify requirements and system properties. Using system simulation techniques at this stage of the development a modification to correct an erroneous execution of the system will be less complicated and require lower costs, since the deployed hardware and software are not yet involved. The early definition and design of i/o interfaces and a late separation between disciplines permit the creation of integrated impact models. Finally, the technical viewpoint contains the implementation, hardware and planning of the system. We consider the viewpoints as development stages with a seamless integration. For instance from the functional view the logical view is derived. From the system implementation models code for the final hardware and execution environment is generated.

4.3 FOCUS Modelling Approach for Mechatronic Systems

In this section, we present our approach to develop mechatronic systems using the FOCUS theory and its implementation, according to the design methodology, presented in the precedent section. Figure 4.2 illustrates the structure of our generalized development approach in a top-down manner. The boxes represent artefacts that have been developed and the arrows show which other artefacts are derived. The process starts by structuring initial requirements using specific syntactic patterns: this first step raises the level of precision by transforming the free text requirements into a structured form using specific pre-defined syntactic patterns, as presented in Fleischmann (2008).

An informal specification consists of a set of words, which can be classified into two categories: content words and keywords. Content words are system-specific words or phrases, e.g., "Off-button is pressed". The set of all content words forms the logical interface of the system with its environment, which can be understood as a special kind of domain specific glossary that must be defined in addition. Keywords are domain-independent and form relationships between the content words (e.g., "if", "then"). Thus, a semi-formal specification consists of a number of requirements described via textual patterns, which can be easily understood even by engineers unfamiliar with formal methods. Using this description to structure the informal specification, missing information can be already discovered. Furthermore, possible synonyms are identified that must be unified before proceeding to a

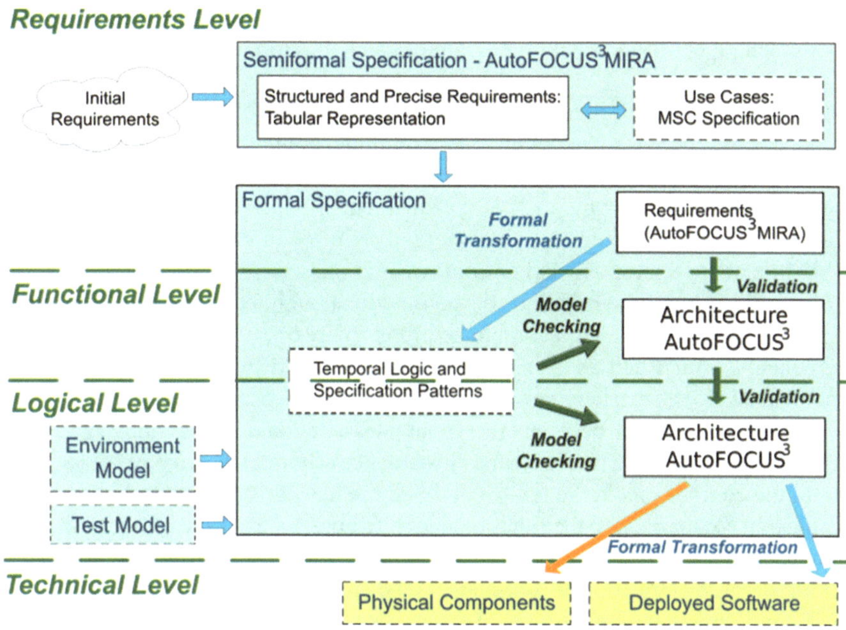

Fig. 4.2 A representation of our generalized modelling approach

formal specification. Analysis of the semiformal specification document should also detect sentences, which need to be reformulated or extended. This specification can now be schematically transformed to a Message Sequence Charts (MSCs) representation, as an optional step relevant for highly interactive systems. Our approach for the development of embedded system was already used in industrial case studies (Campetelli and Spichkova 2012).

The methodology proceeds with the translation of semi-formal specification to AutoFOCUS 3. As mentioned above, AutoFOCUS 3 is a modelling tool based on the semantics and the model of computation of the FOCUS modelling theory. AutoFOCUS 3 supports the development of reactive, software-intensive, embedded systems and is implemented on top of the Eclipse[3] platform. In this tool, systems are modelled by software architectures composed of components with executable behaviour descriptions. AutoFOCUS 3 supports timed synchronous components with a discrete notion of time, that is, a subdivision of time in logical ticks or steps, in which the model components synchronously interact according to global clocks. Development views in the tool support viewpoints from the SPES matrix. A requirement framework called Model-based Requirements Analysis (MIRA) supports the specification of system requirements informally guided by templates (Teufl et al. 2013). The following formalization step of these informal

[3]http://www.eclipse.org

specifications is done using a set of integrated formal notations. This way the requirement specifications show a complete and structured description of system behaviour.

From the formal requirements is derived a hierarchy of system functions and sub-functions, and their behaviour. The requirements interface guide the mapping from requirements to functions and become part of the interface of the functions. The functional architecture of the system is then a set of functions and their relations. The functional model is given by a set of communicating systems, each one with a defined interface and an implementation.

The realization of the functions of the functional architecture is modelled in the logical architecture. The logical architecture is a network of interconnected logical components, which can be also hierarchical structured in subcomponents. Each component has an interface composed by a set of i/o ports and its behaviour is specified by a relation between the input messages and output messages. The components exchange typed messages instantaneously through i/o ports. The interface of the components in the logical architecture has ports and types that are more technical in comparison to the functional architecture. One of goal of our modelling approach is to obtain comprehensive functional and logical architectures of the system. In these architectures, a fundamental step is to determine the context of the system and the relevant properties of the system environment.

We model with AutoFOCUS 3 two kinds of specifications: a formal specification of system requirements and corresponding architecture specifications. This prepares the basis to verify the system architecture specifications against the requirements using model checking techniques. The requirements specification is schematically translated to temporal logic or specification patterns, which gives basis to model-check the model (Campetelli et al. 2011). Model checking in AutoFOCUS 3 supports the following features: tight coupling of verification properties with model elements, visualization and simulation of counterexamples, and different specification languages for the formulation of properties. Formalized requirements are checked against the functional and logical architectures with automated analysis techniques, as simulation, formal verification, and model-based testing.

Finally, we proceed from the logical to the technical level, where we split our model into software and physical components. Hardware aspects are captured in a topology model, which describes execution and transmission units such as electronic control units and bus systems. A deployment model allocates components to execution units and allows generating C code of the system, which can be compiled and installed into the demonstration hardware. We have shown that the C program produced by the AutoFOCUS 3 code generator is a reasonable simulation of the model. Altogether, the methodology guides us from an informal specification via stepwise refinement to a verified formal specification, a corresponding executable verification model, and a C code implementation.

4.4 FOCUS Modelling Foundations for Mechatronic Systems

In this section, we describe in more detail the FOCUS modelling theory for the specification of the logical representation of mechatronic systems, according to the presented methodology. We can model a system in FOCUS, beginning at an abstract requirement specification, which can be formalised for instance as a functional specification. These specifications represent the foundation for the following phase, which is a concrete implementation description. As described above FOCUS models are structured as hierarchical components connected with input and output channels. An example of the hierarchical and interconnected net of components as defined in FOCUS, each with a typed i/o interface, is depicted in Fig. 4.3.

A system is composed of a number of subsystems (called its components). The composed system is a component itself and can in turn be a part of a larger system. A component is specified by its interface to communicate with its environment and an encapsulated state define a component. A component has typed input $I = \{x_1 : T_1, x_2 : T_2, \ldots\}$ and output $O = \{y_1 : T_1', y_2 : T_2', \ldots\}$ channels. We denote with $(I \square O)$ the syntactic interface of the component. Infinite and finite sequences of elements from given sets are called streams. Streams can consist of actions (called traces) or of messages (called communication histories). A data stream function $x : N_+ \rightarrow T^*$ describes the behaviour of a channel of type T. The interface behaviour of a component with syntactic interface $(I \square O)$ is defined with $[I \square O] = \{H[I] \rightarrow \wp(H[O])\}$ that

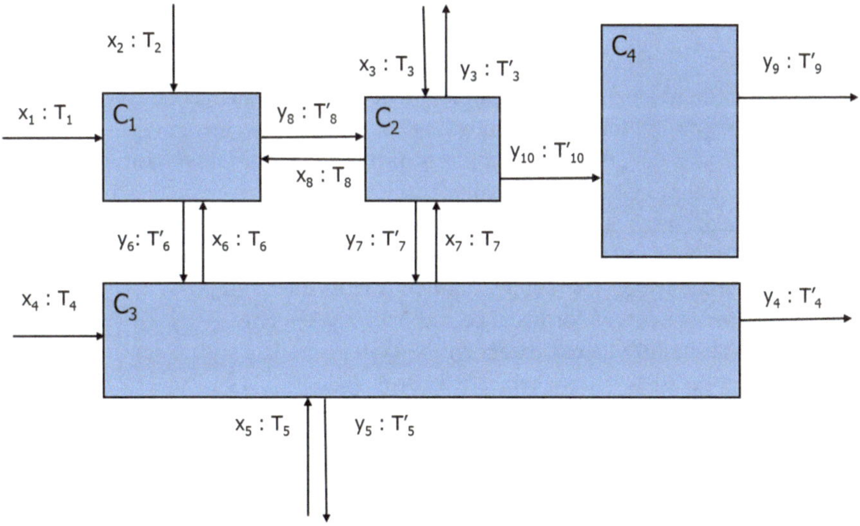

Fig. 4.3 An example of FOCUS logical architecture with interconnected components. Components have input channels (x_i) and the output channels (y_j) that have respectively T_i and T'_j type

is the set of all component executions. Data streams are a central concept in FOCUS theory.

The internal behaviour of each component can be implemented using different formalisms, for instance functional specifications or finite state machines. A single state represents a snapshot of the system, and through the actions the system processes from a state to the next, determining an evolution of the system from one snapshot to the following.

Definition 1 (Transition System)
A transition system is a tuple $(Act, State, \rightarrow Init)$ where:

Act is a set of actions,
$State$ is a set of states,
$\rightarrow \subseteq State \times Act \times State$ is the transition relation,
$Init \subseteq State$ is the set of initial states.

The transition $(\sigma, a, \sigma') \in \rightarrow$ is also written as $\sigma \xrightarrow{a} \sigma'$. The execution of a transition system are sequences of states and actions: $\sigma_0 \xrightarrow{a_1} \sigma_1 \xrightarrow{a_2} \sigma_2 \ldots$ where σ_0 is an initial state and $\sigma_i \xrightarrow{a_{i+1}} \sigma_{i+1}$ holds for all i. The semantics transition system is described by its executions represented sequences of states and actions, while the interface of a system is described by its input and output behaviour. This behaviour is the component interface, which is defined by input and output actions and a predicate on input/output traces. The set of input actions (I) must be disjoint from the set of output actions (O) for distinguishing inputs from outputs in a trace. Predicates are described by trace logic and/or transition systems.

4.5 FOCUS Continuous Time Modelling

At the requirement level, a specification of a distributed system should contemplate a suitable definition for system interface as well as for the environment. The FOCUS approach covers these definitions, providing requirements for the components and assumptions for the environment. For discrete systems, the communication histories or traces are represented by infinite sequences of messages at discrete time intervals. A discrete model of computation can be a limiting restriction for reactive systems, which interact in continuous real-time with physical components and their environment, as for instance mechatronic systems. A recent evolution of the FOCUS modelling theory extended the system models with a continuous time model of computation (Campetelli 2013; Broy 2012).

In order to permit a real-time execution of the system models, time should be represented in the real number realm, that is, the set of all non-negative real numbers. Consequently, the system needs to have continuous data types and an instrument to define the continuous evolution of these data types. We introduce a transition system, inspired by Henzinger's hybrid automaton with a continuous time model: the i/o FOCUS hybrid state machine. With the following definitions, we describe continuous time streams and the syntactic interface of a FOCUS component that implements them, called FOCUS hybrid component.

Definition 2 (Hybrid Stream)
Let M be the set of all messages (potentially infinite), a hybrid discrete stream x over M is described by a function:

$$x : I \to M^*, \text{ with } I \subseteq \mathbf{R}_+$$

whereas a hybrid continuous stream y over a set N of messages (typically $N = R_+$) is described by a total function:

$$y : \mathbf{R}_+ \to N$$

Definition 3 (Syntactic Interface)
The syntactic interface for FOCUS hybrid components is a function with (m + k input and n + l output), where each element is a hybrid stream:

$$h : \left(M_1^*\right)^{R_+} \times \ldots \times \left(M_m^*\right)^{R_+} \times (L_1)^{R_+} \times \ldots \times (L_k)^{R_+} \to \wp\left(\left(N_1^*\right)^{R_+} \times \ldots \times \left(N_n^*\right)^{R_+} \times (O_1)^{R_+} \times \ldots \times (M_l)^{R_+}\right)$$

where m may be equal to zero if there are discrete streams in the input, also n may be equal to zero if there are no discrete streams in the output. k may be equal to zero that means no input continuous streams, also l may be equal to zero that means no output continuous streams. Anyway, n and l cannot be both at the same time equal to zero. The sets of messages are not necessarily different.

The behaviour of hybrid components can be deterministic or nondeterministic. It is deterministic if the function h returns only one output sequence for each input sequence. We define the implementation of a hybrid component based on the definition of hybrid automata and our notion of component in FOCUS.

Definition 4 (I/O FOCUS Hybrid State Machine)
An i/o FOCUS hybrid state machine is a tuple $H = (\Sigma, Var, Init, I, O, Dom, E, f, G, R)$ with:

- A state space $\sum = (Q \times V)$ where Q is a set of discrete states $Q = \{q_1, q_2, \ldots\}$ and V a set of continuous states $V \subseteq \left(M_1^*\right)^{R_+} \times \ldots \times \left(M_l^*\right)^{R_+} \times (M_{l+1})^{R_+} \times \ldots \times (M_n)^{R_+}$, where n is the total number of variables. To each element of V a variable in the set Var is associated.
- A set of initial states $Init \subseteq \Sigma$.
- A set of input $I \subseteq V$ and output $O \subseteq V$ states, respectively for input and output channels, with $O \neq \emptyset$. To each element of I and O corresponds a hybrid stream. The set $Int \subseteq V$ are the internal continuous states and $Int_d \subseteq V$ the internal discrete states both with no streams associated. Any variable in Var can only

be in one of these sets. $O_d \subseteq O$ and $O_c \subseteq O$, with $O_c \cap O_d = \emptyset$ are the continuous and discrete output states respectively, we denote with $W = Int \cup O_c$ the set of internal and output states.

- A domain function $Dom : Q \rightarrow \wp(V)$.
- A set of edges $E \subseteq Q \times Q$ that represent the discrete state transition.
- A vector field function $f : Q \times V \rightarrow W$.
- A guard condition function $G : E \rightarrow \wp(V)$.
- A reset map function $R : E \times V \rightarrow \wp(W \cup O_d)$ that resets the variables at a discrete state transition.

A hybrid state i/o machine in FOCUS communicates through input and output channels over hybrid streams. The continuous state space V is subdivided into internal variables, output and input subsets. The i/o variable subsets are associated with i/o channels of the component. In each discrete state, the vector field function describes the evolution of the continuous variables, which e.g. are guided by differential equations. Differential equations are widely used to describe the logical behaviour of mechatronic systems that work with physical or mechanical parts. The transitions between the discrete states are decided by the guard function. The execution of the i/o FOCUS hybrid state machine by a sequence of continuous and discrete modifications is characterised by the state transitions, and are influenced by the input values received through the input streams.

The parallel composition of two or more hybrid components forms a net of components, which can be represented by a directed graph. In this graph, the components are nodes and the edges correspond to communication channels. Our composition of i/o hybrid state machines is based on the same principle as the well-known composition of state machines.

4.6 Simulation of I/O FOCUS Hybrid State Machines

We introduced a continuous time model of computation to execute the I/O hybrid state machines. In our design methodology, software-modelling tools without specialized hardware are used for the logical representation of the system. In such tools, a computable simulation is executed using a discretization of the differential equations in the models through sampling techniques. In communication/signal theory, the term sampling indicates the operation to approximate an analogue signal with a discrete signal. Sampling is a consolidated and widely used solution to discretize continuous signals, whereas numerical analysis provides the theory to formalize the sampling of variables associated to differential equations. The discrete elaboration steps must be as large as possible to guarantee a good level of performance of the digital simulation, and at the same time as short as possible to guarantee a desirable precision of the results. The time between a sampled variable value and the next value is called period. The sampling may be periodic if the sampling period is constant and variable if the period is not constant. Sampling with a variable period is a common approach, called adaptive step size control, which guarantees a desired accuracy of the solution while providing reasonable computational time. The additional computation for the adaptation of the period is compensated by the overall advantages of the adaptive solutions.

We build an approach inspired by the work in Petreczky et al. (2009), however, our algorithms have a variable sampling step size. We approximate the differential equations in the I/O Focus hybrid state machines attempting to reduce the complexity together with the desired level of accuracy. Variable step solutions are widely used in industrial tools; they vary the time step according to the size of the predicted error of the approximated differential equations and predefined constant values. Widrow and Hoff (1989) proposed one of the fundamental variable step sampling approaches.

We studied and elaborated two different solutions for adjusting the change of the period length (Campetelli 2013; Campetelli and Hackenberg 2015). The first algorithm is based on the gradient calculated between the actual and the precedent value of the continuous variables; and the second one is based on predefined time intervals for the values of the variables, which have associated sampling periods. The sampling architecture produces sequences of sampled values from i/o continuous streams. We define sequences of sampled values from a hybrid stream.

Definition 5 (Sampled Hybrid Continuous Stream)
Considering a finite set M where $\perp \notin M$ and a finite or infinite timed sequence of elements s of M:

$$s = (m_1, t_1)(m_2, t_2)\ldots(m_k, t_k)\ldots$$

where $0 \leq t_1 < t_2 < \ldots < t_k < \ldots$, $m_{i+1} \in M$, $t_{i+1} \in \mathbf{R}_+$ for $i \in \mathbf{N}$, $i < |s|$. We associate a hybrid continuous stream α to a sampled sequence:

$$\alpha : t \in \mathrm{R}_+ \mapsto \begin{cases} m_{i+1} \in M & \text{if } t = t_{i+1} \text{ for some } i \in \mathrm{N} \\ \perp & \text{otherwise} \end{cases}$$

In an analogue manner a sampled hybrid discrete stream can be defined.

Now we describe the sampling algorithms in detail. In both solutions, the sampling period is initially set to a predefined value. The first solution considers the gradient between the actual and the precedent value of the continuous variables. The gradient is calculated between the actual value and the value of the precedent elaboration step of the variable. If the gradient is smaller or equal to a predefined acceptance value then the period remains constant, otherwise is set to half. In the same way if in a predefined stabilization time the gradient remains smaller or equal to the acceptance value and the period is less than a maximum value, then the period is doubled, eventually until a maximum value is reached. Acceptance and stabilization time values are defined manually before the simulation according to the differential equations in the model. This way the sampling is effective and has a desirable precision. We have some preliminary ideas to determine these values semi-automatically with formal verification procedures. The second solution is based on continuous variable value intervals, called critical intervals, each with a corresponding period value. When a variable is in a critical interval then the period must be equal or smaller to a correspondent value, called acceptance period, otherwise the period is set to it. In both solutions, the algorithm is applied to each continuous variable and then the smallest necessary period is chosen.

MATLAB/Simulink[4] is a toolbox that represents the state of the art for modelling and simulation of mechatronic systems. We implemented our algorithms in MATLAB and performed preliminary tests respect to the solutions based on the estimation of the error of the approximated values of the variables included in the tool. In these tests our solutions required less computational time, which is important for the scalability of complex scenarios. Another advantage comes from the control of the precision of the simulation in the predefined value intervals. This way we obtain a tailored simulation, optimized for validation and analysis purposes, instead for the precision of its results.

4.7 Formal Verification of FOCUS Models

Mechatronic systems are part of highly safety-critical systems such as control systems for vehicles, machines aircraft or medical instruments. Verification is a crucial aspect of their development, but also challenging due to its complexity. One prominent problem is that verification cannot easily handle the state space dimension of mechatronic systems, whose size is determined also by the number of continuous variables. Testing and simulation are widely used validation techniques. These techniques only consider a relatively small subset of all possible executions of the system. Compared to (informal) testing, formal verification has the advantage that the verification is made in a complete, semi-automatic or fully automatic exhaustive way, where all the possible executions of the system are considered. On the other hand, there are important issues that may reduce its applicability, e.g. not optimal integration in modelling tools, skills that are required to use it, enough time or memory resources for some verifications.

Despite its complexity, formal verification tools have been introduced in industrial development projects. There are two fundamental verification techniques: model checking and theorem proving. For these techniques: first, a mathematical model of the system, and second, formally specified requirements, which the model should satisfy, have to be provided.

We believe in better-integrated development environments, where design and verification tasks are strictly linked together with faster and more usable methods. In this integration, also the usability and the integration in support tools are important aspects, because the effectiveness of powerful verification solutions

[4]http://www.mathworks.com

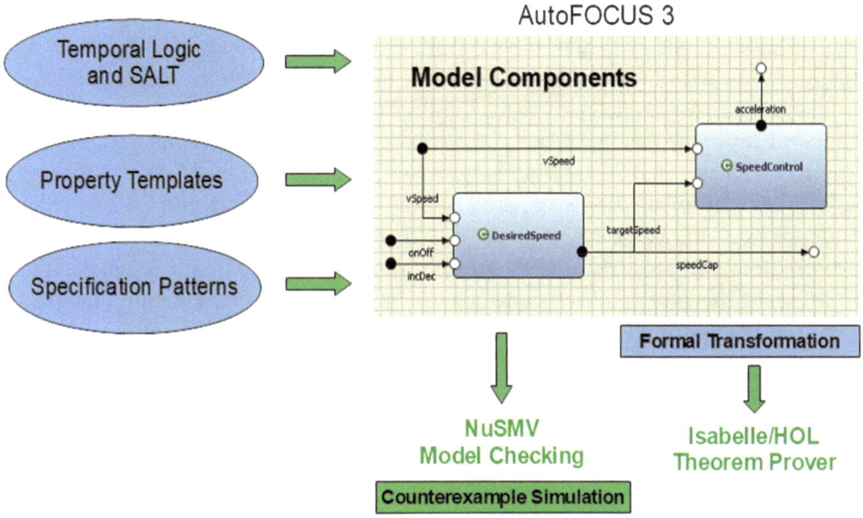

Fig. 4.4 Formal verification approach in the tool AutoFOCUS 3

may be invalidated by a not optimal integration, or by the high skills required to manage the tools and the verification properties.

Our modelling tool AutoFOCUS 3 provides an interactive graphical simulation environment and testing/verification capabilities for the logical architecture. Formal verification integration in AutoFOCUS 3 is depicted in Fig. 4.4.

We built a user-friendly integration of the model checker NuSMV[5] in the modelling environment. The choice of NuSMV as model checker is mainly due to its semantics. In fact, in AutoFOCUS 3 the interconnected components execute an elaboration step synchronously, in the same manner as the modules in a NuSMV model. Furthermore, the symbolic model checking provided by NuSMV works well with hardware-like systems. NuSMV guarantees one of the best performances for the formal verification of such systems available (Cimatti et al. 1999). For the execution of the NuSMV model checker, AutoFOCUS 3 automatically translates the selected component with all its subcomponents into an SMV instance. We performed preliminary invariant verifications of i/o FOCUS hybrid state machines with a special version of NuSMV for hybrid systems (HyCOMP[6]). The verification of invariants of continuous variables can be used in future for the automatic determination of parameters for the introduced sampling algorithms.

We demonstrate applicability of verification mainly in three areas. First, we integrate specification and verification tightly into the model-based development

[5]http://nusmv.fbk.eu

[6]https://es-static.fbk.eu/tools/hycomp

process. This means that verification properties are linked to model elements and can be saved along with the model itself. The properties can be verified locally easily during the development. In addition, the support for different specification languages is essential. Properties, as highlighted in Fig. 4.4, can be expressed with temporal logic, as for instance Linear Temporal Logic and Computational Tree Logic. The Structured Assertion Language for Temporal Logic (SALT) provides a higher level of abstraction compared to other temporal logics formalisms (Bauer et al. 2006). This is based on ideas of existing approaches, such as specification patterns but also provides nested scopes, exceptions, support for regular expressions, real-time, and employs some constructs similar to a programming language. The second approach is a pattern-based approach for the presentation, codification and reuse of property specifications for finite-state verification. Specification patterns permit to describe properties for the model checker at a high level of abstraction (Dwyer et al. 1999). Patterns are generalized description of occurring requirements concerning aspects of the system's behaviour. The behaviour of the system is modelled as state/event sequences in a finite-state model. We added most of the specification patterns in the model checker view of AutoFOCUS 3, where parts of the patterns can be defined and customized with logical operators and elements of the models. The third type of specifications are specific properties templates that can be directly selected, configured and executed from the graphical interface.

We implemented property templates for the simpler but recurrent cases and high-level languages for the more complex properties. Properties disproved by verification must be analysed and either the system model or the properties have to be corrected. NuSMV checks whether the system satisfies a property, and provides a "yes" or "no" answer. If the system does not satisfy a property, the answer is "no" and a counterexample is provided, i.e., a trace (system run) violating the property. In AutoFOCUS 3 counterexamples can be either simulated or represented as a message sequence chart. AutoFOCUS 3 models can be verified with the theorem prover Isabelle/HOL[7] through a formal model transformation (Spichkova 2007). In practice, theorem proving expresses the property and the system in mathematical logic as a set of axioms and a set of inference rules; and finds a proof of a property from the axioms. The proof is composed of steps, which invoke the axioms and rules, and derive definitions and intermediate lemmas if possible. Model checking is completely automatic in contrast to theorem proving. In AutoFOCUS 3, system models are encoded in Isabelle/HOL, as well as proof of theorems that support subsequent verification of properties in Isabelle/HOL.

[7]http://www.cl.cam.ac.uk/research/hvg/Isabelle

4.8 Conclusions

The design of mechatronic systems has to face complex domain requirements, multidisciplinary and interdisciplinary issues, and a strict link between the technical level and the hardware implementation. From that the needs for a seamless, agile and efficient model-based development methodology follows, which guarantees also adequate and rigorous quality control procedures. We described a modelling approach founded on a formal modelling theory, with a methodology for the design and handling of mechatronic systems. The methodology comprises seamless artefact-based modelling, simulation, and formal verification aspects. The SPES modelling framework permits a structured and systematic development focused on system artefacts in four viewpoints, namely the requirements, functional, logical and technical viewpoints.

The presented formal modelling approach proposes a logical model representation that can be used by different experts involved in the design of mechatronic systems, such as computer scientists, electrical engineers, and mechanical engineers. The logical representation describes the behaviour of the system in a structure with typed interfaces and is derived from the requirements and functional artefacts. The abstraction of technical details in the logical representation ensures less complex integration between the involved disciplines models, which can be simulated, verified and validated. The verification of requirements and properties is executed without the necessity to have the final hardware. As consequence, changes in the logical models can be implemented more easily and cheaper than the same change when discovered at the technical representation. Formal verification enhances the reliability of the system by ensuring in an exhaustive manner that the model meets its functional requirements, before the system is implemented. This way the development life cycle is more effective as bugs are detected earlier. We are aware of the intrinsic difficulties in a synthesis of a suitable modelling theory with formal verification capabilities, but the overall advantages can outweigh the efforts necessary to modify the existing design methodologies.

High-automated tasks of mechatronic systems interact with e/e parts, physical components, and the environment. Some systems can be represented by discrete time components because the continuous dynamic of the system is abstracted. However, there are systems, especially safety-critical systems that require models with differential equations and continuous time simulation capabilities for their logical representations. Therefore, we presented recent extensions to the FOCUS modelling theory to support a continuous time model of computation. We want to reach a more precise and better representation of mechatronic systems, together with the streams theories, the abstraction and modularity of components available in FOCUS. Since we believe that the simulation of hybrid i/o state machines in FOCUS should support validation purposes and not only the precision of the approximation, we also introduced dynamic sampling solutions.

Further evolutions of the formal modelling theory are for instance the introduction of product lines or instantiation of predefined modules/components, and the support for specific energy, element or material flows. Case studies with feedback from different discipline experts, to prove our approach and debate further developments remain a topic for future research projects with industrial partners. An implementation of the hybrid components and their dynamic sampling and verification in AutoFOCUS 3 is also part of future work.

Altogether, the methodology guides us from informal specifications via stepwise refinement to a verified formal specification, a corresponding executable verification model, and a C code implementation. The proposed modelling approach can be considered as a first step towards an integrated interdisciplinary design of mechatronic systems.

References

Anacker, H., Dorociak, R., Dumitrescu, R., Gausemeier, J.: Integrated tool-based approach for the conceptual design of advanced mechatronic systems. In: 5th Annual IEEE International Systems Conference, pp. 506–511. IEEE, Montreal (2011)

Bauer, A., Leucker, M., Streit, J.: SALT – Structured assertion language for temporal logic. Proceedings of the Eighth International Conference on Formal Engineering Methods, pp. 757–775 (2006)

Becker, S., Dziwok, S., Gerking, C., Heinzemann, C., Thiele, S., Schäfer, W., Tichy, M.: The MechatronicUML Design Method – Process and Language for Platform-Independent Modeling. Heinz Nixdorf Institute University of Paderborn, Paderborn (2014)

Broy, M.: System behaviour models with discrete and dense time. In: Advances in Real-Time Systems, pp. 3–25. Springer, Berlin (2012)

Broy, M., Dederich, F., Dendorfer, C., Fuchs, M., Gritzner, T., Weber, R: The design of distributed systems – an introduction to FOCUS. Technical Report TUM-I9202. Technische Universität München (1992)

Campetelli, A.: Dynamic sampling for FOCUS hybrid components. In: Proceedings of the 3rd International Conference on Circuits, System and Simulation (ICCSS'13). Int. J. Model. Optim. 402–406(2013)

Campetelli, A., Hackenberg, G.: Performance Analysis of Adaptive Runge-Kutta Methods in Region of Interest. In: 2nd International IFIP Workshop on Emerging Ideas and Trends in Engineering of Cyber-Physical Systems, Dresden, Germany (2015)

Campetelli, A., Spichkova, M.: Towards system development methodologies: From software to cyber-physical domain. In: Proceedings of First International Workshop on Formal Techniques for Safety-Critical Systems. Open Publishing Association. https://arxiv.org/pdf/1403.2819.pdf (2012)

Campetelli, A., Hölzl, F., Neubeck, P.: User-friendly model checking integration in model-based development. In: 24th International Conference on Computer Applications in Industry and Engineering (2011)

Cimatti, A., Clarke, E.M., Giunchiglia, F., Roveri, M.: NUSMV: a new symbolic model verifier. In: Computer Aided Verification, pp. 495–499. Springer, Trento (1999)

Dwyer, M.B., Avrunin, G.S., Corbett, J.C.: Patterns in property specifications for finite-state verification. In: Proceedings of the 21st International Conference on Software Engineering, pp. 411–420. ACM, New York (1999)

El-khoury, J., Redell, O., Törngren, M.: A tool integration platform for multi-disciplinary development. In: 31st EUROMICRO Conference on Software Engineering and Advanced Applications, pp. 442–450. IEEE, Porto, Portugal (2005)

Fleischmann, A.: Model-based formalization of requirements of embedded automotive systems. TU München (2008)
Habib, M.: Mechatronics a unifying interdisciplinary and intelligent engineering science paradigm. IEEE Industrial Electronics Magazine (2007)
Henzinger, T.A.: The theory of hybrid automata. In: 11th Annual IEEE Symposium on Logic in Computer, pp. 278–292. IEEE Computer Society, New Brunswick, NJ (1996)
Kernschmidt, K., Vogel-Heuser, B.: An interdisciplinary SysML based modeling approach for analyzing change influences in production plants to support the engineering. In: 9th IEEE International Conference on Automation Science and Engineering, pp. 1113–1118. IEEE, Madison, WI (2013)
Petreczky, M., Beek, D.A., Rooda, J.E., Collins, P.J., Schuppen, J.H.: Sampled-data control of hybrid systems with discrete inputs and outputs. In: Proceeding of the 3rd IFAC Conference on Analysis and Design of Hybrid Systems. International Federation of Automatic Control (2009)
Pohl, K., Hönninger, H., Achatz, R., Broy, M.: Model-Based Engineering of Embedded Systems – The SPES 2020 Methodology. Springer, Heidelberg (2012)
Schäfer, W., Wehrheim, H.: The challenges of building advanced mechatronic systems. In: 29th Intertational Conference on Software Engineering – Future of Software Engineering, pp. 72–84. IEEE Computer Society, Minneapolis, MN (2007)
Spichkova, M.: Specification and seamless verification of embedded real-time systems: FOCUS on Isabelle. Phd dissertation. Technische Universität München (2007)
Teufl, S., Mou, D., Ratiu, D.: MIRA: a tooling-framework to experiment with model-based requirements engineering. In: Proceeding of the 21st Requirements Engineering Conference (RE), pp. 330–331. IEEE International (2013)
Thramboulidis, K.: The 3+1 SysML view-model in model integrated mechatronics. IEEE Transactions on Industrial Informatics, pp. 109–118. IEEE International (2010)
Widrow, B., Hoff, M.: Adaptive switching circuits. In: WESCON Conference Record. (1989)

Chapter 5
Functional System Architecture for an Autonomous on-Road Motor Vehicle

Richard Matthaei and Markus Maurer

Abstract Autonomous driving is a widely discussed field of research with still growing interest. In addition to a lot of technical, legal and social questions to be solved, an immense challenge still remains in mastering the complexity of the resulting system which would eventually replace the driver. A supporting tool for developing complex systems is given by the functional system architecture, which describes the system on an abstract level independent of concrete technical solutions. Functional system architectures published in the context of autonomous driving do not cover all necessary functional requirements. However, they focus on different sub-aspects and functional mechanisms within this context.

Our functional system architecture, which has been developed in the research project Stadtpilot at the Technische Universität Braunschweig, focuses on systematization and a combination of localization- and perception-driven approaches into one single well-structured functional system architecture. It has been developed in a top-down approach based on a formulation of the functional requirements of an autonomous on-road motor vehicle, in the sense of a modular building block system. It covers the aspects of localization, environmental and self-perception, mission accomplishment, usage of map data and communication, and the integration of the human being as a passenger and as another traffic participant in the close surroundings of the autonomous vehicle.

Referring to our functional system architecture, we discuss some basic mechanisms of autonomous driving in the following article, which become transparent due to the architecture's basic structure. Additionally, we discuss where current advanced driver assistance systems are located within this architecture. This makes the big efforts which still have to be made to fulfill the necessary functional requirements regarding an autonomous vehicle driving safely in public road traffic more transparent.

Keywords System architecture • Autonomous driving • Localization • Map data • Perception • Cooperation

R. Matthaei (✉) • M. Maurer
Institut für Regelungstechnik, Technische Universität Braunschweig, Hans-Sommer-Straße 66, Braunschweig, 38106, Germany
e-mail: richard.matthaei@t-online.de

H. Winner et al. (eds.), *Automotive Systems Engineering II*,
DOI 10.1007/978-3-319-61607-0_5

5.1 Introduction

The vision of "autonomous driving" is widely discussed nowadays. The estimates of an introduction into the market diverge significantly. All options from 5 to 20 years, or even "never," are mentioned. This high divergence is conspicuous and might be caused by various reasons. One reason we identified is a very heterogeneous understanding of the capabilities of an "autonomous" or fully automated vehicle. The use case definitions vary from just following a certain lane on a highway without any lane changes (this is discussed, for example, in Chap. 6) up to a fully automated taxi traveling in crowded urban environments.

It becomes clear that there is a big gap between these two use cases concerning the functionality of an "autonomous" vehicle. This gap is currently not considered by the definitions of Gasser et al. (2012), but it is already part of the SAE (Society of Automotive Engineers) levels (SAE International 2016). In the case of full automation according to the SAE level five "full automation" (SAE International 2016), the abilities of the vehicle equal the vision of a fully automated taxi.

This vision is the basis for the following discussion. It is a top-level use case in our understanding, because it covers all relevant sub-use cases in the context of driving a vehicle in public road traffic. A definition of the functional requirements for this use case and the state of research into functional system architectures is discussed in Matthaei and Maurer (2015).

In order to gain a better understanding of the system of an autonomous vehicle and also to master the growing complexity of the systems for vehicle automation, we designed a functional system architecture (Matthaei 2015; Matthaei and Maurer 2015) which covers all aspects summarized in Sect. 5.3. In the sense of a modular building block system, our functional system architecture also supports a structured development of (sub-)systems, on the one hand, and allows the comparison of existing approaches to this holistic architecture, for example, to get an idea of the remaining functional gap towards autonomous driving, on the other hand.

5.2 What is a Functional System Architecture?

A great challenge while developing systems such as autonomous vehicles is the handling of an immensely complex system. In our special case, this also includes bringing various existing approaches together into one scalable system. A well-known approach to manage this challenge is a step by step subdivision of the entire task into smaller subtasks. The resulting functional modules and their dependencies (interfaces) are then described in a so-called logical or *functional system architecture*. The step from a functional system architecture to a technical system architecture is carried out by selecting a certain technical solution for a functional task, such as using a Kalman filter (technical solution) for estimating dynamic environmental features (functional task).

As mentioned already, the main objective concerning the development of a functional system architecture is managing the complexity. The resulting system should be testable, maintainable and scalable. This includes, in addition to the definition of modules, their interfaces, their functionalities and their dependencies, also the organization of the team and the project management (e.g. definition of a road map). A functional system architecture, which is accepted or even developed by the entire team, supports the discussion within the team as it defines the central terms and a rough structure of the system. Both are fundamental to a common understanding of the system among the team members. Once a functional system architecture is developed, it is possible to identify the vacant functional modules, unintended loops within the information flow or aspects of functional redundancies. Additionally, sub-modules, such as a lane-keeping system, can be developed having the entire system already in mind, so that integrating it into a more complex system later on is possible without completely redeveloping that specific sub-module.

In order to somehow judge the quality of a functional system we tried to identify some aspects which indicate whether a functional system architecture is useful or not. Some of these aspects are already mentioned in the ISO 26262 (2011, pp. 10–13): It claims modularity, a certain granularity and simplicity. Simplicity includes, according to our understanding, a clear information flow, minimization of inter-module dependencies and a limited number of elements/modules. Additionally, the architecture should be presentable and extendable in such a way that new requirements should not lead to a complete redesign, but should be integrated only by minor changes and, of course, it should be complete in a sense that all requirements are covered.

In general, there are mainly two ways of creating an explicit functional overview in the form of a functional system architecture. Systems are often built up by composing more and more sub-modules and new functionalities step by step, especially in young fields of research or development. The progress is then documented by creating an image of the entire system. This procedure would somehow correspond to a *bottom-up approach* and contains the risk of developing towards a dead end. The other way is a more stringent development starting at the system requirements, commonly known as a *top-down approach*. In a first step, a functional system architecture of the item (according to ISO 26262 2011) is designed and checked against the requirements in an iterative process. Once all the requirements seem to be covered by the architecture, the further development of the system is carried out (definition of technical, software and hardware architectures). A "proof" whether a certain functional system architecture is really suitable for a certain task can only be given by the long-term stability of the basic structure over several real-world implementations.

The system architecture we proposed in Matthaei and Maurer (2015) and which is discussed in more detail within this article is developed based on various existing system architectures and checked against the functional requirements of an autonomous on-road motor vehicle. Thus, it uses the experiences of various bottom-up approaches and now follows the top-down strategy for system development. In the next section, the functional requirements of an autonomous on-road motor vehicle are summarized.

5.3 Aspects of Autonomous Driving

We derived the following aspects which are relevant for an autonomous vehicle based on the functional requirements in Matthaei et al. (2015), which are, in turn, based on the remarks in Wachenfeld et al. (2015):

- *Operating*: The vehicle has to be instructed by a human being.
- *Mission accomplishment*: The vehicle has to accomplish the mission defined by a human being. This includes the navigation task, the behavior generation and the control of the actuators.
- *Map data*: Map data is required for route planning purposes in particular. Automated map updates have to be considered.
- *Localization*: The vehicle needs to know its global pose for the usage of map data (e.g. navigation tasks) and communication purposes (e.g. vehicle-to-vehicle (V2 V) or vehicle-to-infrastructure (V2I) communication).
- *Environmental perception*: The vehicle has to perceive its local stationary and mobile environment, including the dynamics of the mobile elements.
- *Cooperation*: The vehicle has to react to the intentions of other traffic participants (automated vehicles and human drivers) and it has to communicate its own intentions to other traffic participants.
- *Safety*: It must be ensured that the vehicle does not constitute any danger above an accepted level to its environment.
- *Self-perception*: The vehicle needs to monitor its current state (functional capabilities of its components, motion, etc.).

The aspects of interior surveillance, as well as aspects of security concerning misuse and manipulation, are not discussed within this article.

5.4 Functional System Architecture

The goal of the proposed functional system architecture is to provide a modular building block system which considers the aforementioned aspects. It is, however, not necessary to implement all of the blocks identified to develop a running system. The developer can choose a subset of the modules to design a system according to his/her wishes.

In the context of this architecture, the single vehicle is understood as a part of a superordinate system. The architecture developed combines a subset of elements of published architectures having an inner-city intersection assistant (as an example of an advanced driver assistant system for complex use cases) and autonomous driving (as an example of a system with a complex functionality) in mind.

This functional system architecture is shown in Fig. 5.1. It is designed as a hybrid architecture including the advantages of a sequential sense-model-plan-act and a parallel behavioral architecture. The main structure of the system architecture

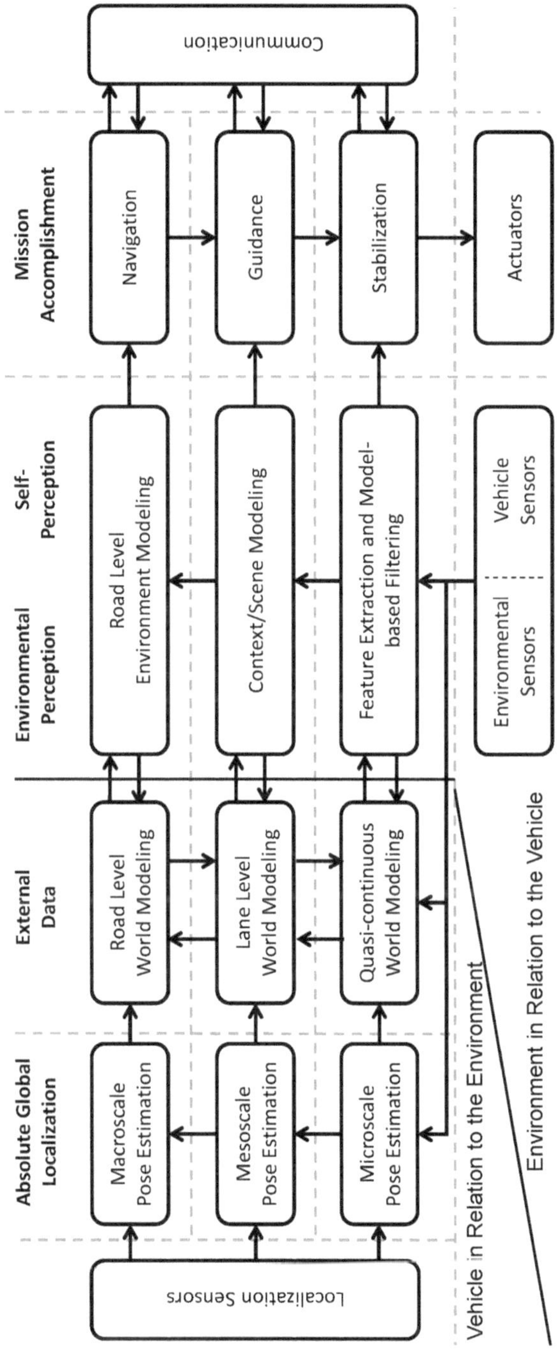

Fig. 5.1 Functional system architecture for an autonomous on-road motor vehicle in public road traffic according to Matthaei and Maurer (2015)

is a three-level design similar to the multilevel designs of Bonasso et al. (1997), Donges (1999), Du et al. (2004) and Maurer (2000), and develops the ideas of Dickmanns (2007) further. Du et al. (2004) also introduced three levels of resolution which are assigned to the aforementioned three levels of the system architecture as follows:

- strategic level: planning, macroscale resolution;
- tactical level: decision making, mesoscale resolution; and
- operational level: reactive stabilization, microscale resolution.

These three levels differ (among other characteristics, as summarized in Tables 5.1 and 5.2) in their resolution, horizon and accuracy (concerning time and space), and relevant environmental features, tasks and cycle times.

Table 5.1 Overview of the different characteristics of the levels, part 1, according to Matthaei (2015)

	Criterion		
Level	Instruction	Task	Environmental features
Strategic	Mission, desired destination	Planning: route planning and navigation	Road network, traffic flow
Tactical	e.g. by emergency vehicles: request for an emergency lane	Deciding: situation assessment and decision unit	Scene, containing the scenery and movable environmental features, their maneuvers, intentions and their context
Operational	e.g. torque demand, brake demand	Executing: vehicle stabilization according to the local environment and within the physical limits	Quasi-continuous image of the environment, exact dynamics

Table 5.2 Overview of the different characteristics of the levels, part 2, according to Matthaei (2015)

	Criterion				
Level	Information of major importance	Localization accuracy, absolute global	Localization accuracy, map	Time-horizon	Horizon
Strategic	Topological	Macroscale (~10–20 m)	Road level	Long-term	From start to destination
Tactical	Semantic	Mesoscale (~1.5 m)	Lane level	Medium-term	Foresighted environment, (~500 m)
Operational	Geometrical	Microscale (quasi-continuous, ~1 cm)	Quasi-continuous	Short-term	Local environment (~100 m)

In an orthogonal direction to these three levels, we introduce the columns "absolute global localization," "external data," "perception" (consisting of environmental perception and self-perception), and the "mission accomplishment." This core of the system (consisting of vehicle and infrastructure) is framed by the sensors, actuators and communication equipment for the exchange of data with human beings or other automated traffic participants.

The columns "perception" and "mission accomplishment" are the state of the research and are already part of many system architectures. They are typically part of a vehicle-referenced view, which means that the environment is described in relation to the vehicle. An absolute global localization is not necessary in this case. The environmental perception focuses on the interpretation of the local environment.

On the contrary, the absolute global localization and the external data describe the overall system "world and vehicle" from another perspective: They describe the environment in an absolute global reference frame, while the global localization determines the pose of the vehicle in relation to this frame. The external data contains information about the stationary and mobile environment, in a sense of a world model and, thus, provides data about the global environment.

A common global reference frame is required, because external data, such as map data, is used by multiple participants. Additionally, the global reference frame is necessary to accomplish the mission in environments without any or with only sparse local features for orientation (e.g. in deserts).

A more detailed discussion of this architecture is published in Matthaei (2015) and Matthaei and Maurer (2015).

5.5 First Findings

Some functional details based on the proposed functional system architecture are discussed in the following sections.

5.5.1 Different Perspectives of the Relation Between Vehicle and Environment

Commonly known system architectures of autonomous vehicles or advanced driver assistance systems treat all sensors as some kind of input to the system. Our architecture differentiates between environmental, vehicle, and localization sensors based on their functional purposes. Additionally, they are connected to the core system on different axis. The environmental and vehicle sensors provide data input to the operational level of the environmental and self-perception. The vehicle sensors acquire data from inside the vehicle (view inwards), whereas the environmental sensors provide data about the vehicle's close surroundings (view outwards). The environmental sensors describe the vehicle's "environment in relation to the vehicle." On the contrary, the localization sensors determine the position of the vehicle within a global reference frame and, thus, describe the vehicle's "position in relation to the world."

These two perspectives describe extreme variants of the fundamentally different approaches of localization-based and perception-based autonomous driving. Perception-based approaches rely more on the perspective "world in relation to the vehicle" (e.g. Bacha et al. 2008; Levinson 2011; Montemerlo et al. 2008; Nothdurft et al. 2011; Rauskolb et al. 2008; Wille et al. 2010; Ziegler et al. 2014), whereas localization-based approaches mainly use the perspective of "vehicle in relation to the world" (e.g. Broggi et al. 2013; Leonard et al. 2008; Müller et al. 2011). We now combined both perspectives within one single system architecture. The localization-based processing is more relevant in the case of missing prominent features in the local environment of the vehicle (e.g. in aviation or nautical applications or in deserts). As soon as local features appear (e.g. obstacles, holes, ways, road markings), the importance of the perception-based processing increases.

This way of designing the system also determines the role of map data. Theoretically, an autonomous vehicle can fulfill its mission (driving collision-free to a desired destination according to the traffic regulations) without map data, relying only on real-time sensor data similar to human abilities (perception-based, blue arrow in Fig. 5.2).

If the stationary environmental features are stored in a local map on the first run, the vehicle can use this data on the way back or on the next run to plan the route and, thus, use the known road network (see Fig. 5.4). No global positioning is required for this theoretical scenario. The technical implementation is theoretically possible using localization mechanisms based on motion sensors (dead reckoning from start point) and simultaneous localization and mapping (SLAM) approaches handling the loop-closure problem (see e.g. Milford and Wyeth 2008; Thrun et al. 2005) to build up drift-free maps.

A common reference system is required in order to share this data with other users. A vehicle would normally use map data which is acquired externally by humans or other vehicles and, thus, can be treated as a priori data. In this case, the a priori map data is a redundant source to the data perceived online for a model of the stationary environment. In the special case of a completely manmade environment, it might also be theoretically possible to develop an automated vehicle which only relies on such map data, V2X communication and absolute global positioning (red arrow in Fig. 5.2).

A parallel redundancy might be achieved by applying both solutions simultaneously.

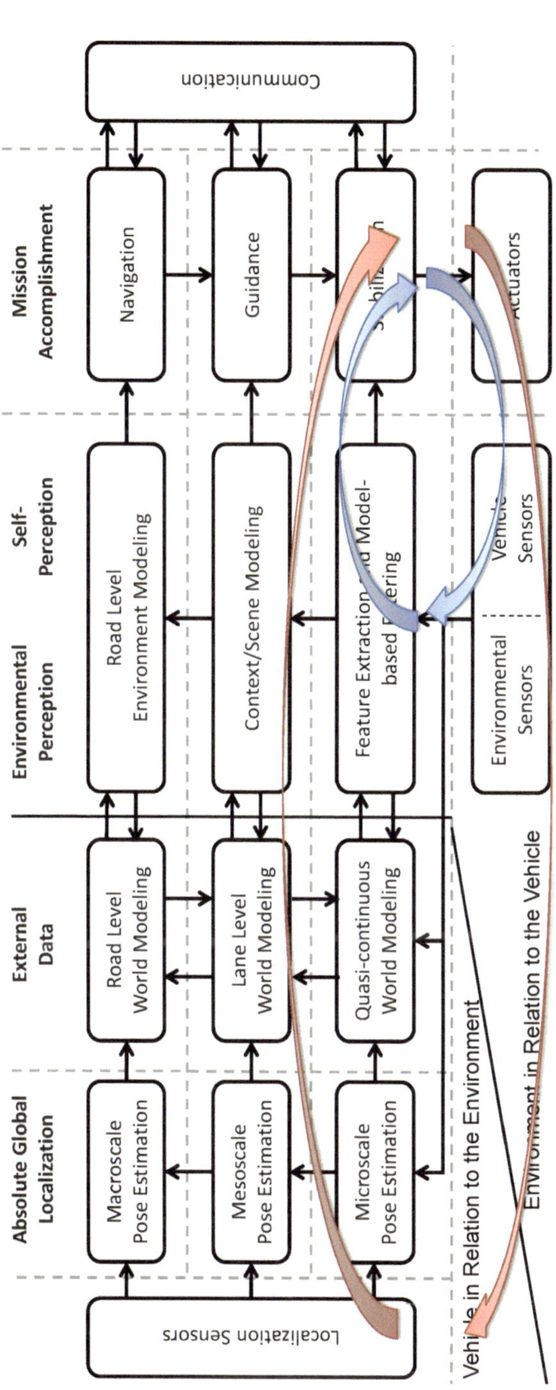

Fig. 5.2 Functional redundancy within the system architecture. *Red loop*: vehicle control based on global navigation satellite systems (GNSS), map data and V2X technologies; *blue loop*: vehicle control based on environmental perception

5.5.2 *Localization Solutions*

The proposed functional system architecture covers a total of four different localization solutions and makes their special tasks transparent.

A full-size system requires two different global localization solutions (see also Fig. 5.3):

- An accurate absolute pose on the world, for example, for automated map updates of the geometries (e.g. the positions of the roads and lanes) on a central backend (see also Visintainer and Darin 2008).
- An accurate map-relative pose for the usage of map data within the system and update of lane or road attributes (see also Visintainer and Darin 2008). This pose can differ from the absolute global pose due to inaccurate map data.

Additionally, two local poses are proposed (see Fig. 5.4):

- A local pose relative to the starting pose of the vehicle, which is mainly applied for motion compensation within the environmental perception. This pose may contain long-term drift, but has to be quasi-continuous regarding successive poses. It might also exist only implicitly due to ego motion compensated processing of sensor data, which results in an egocentric view of the environment.
- A map-relative local pose, as far as no global pose is available, for mapping purposes. This pose has to be drift-free, but need not necessarily provide a continuous sequence of pose.

This concept subdivides the proposed global localization of Moore et al. (2009) into three independent localization solutions (global, global map-relative and local map-relative pose) and keeps the postulated local pose as a localization solution which has to remain consistent to a motion estimation of the vehicle.

5.5.3 *Prediction of the Dynamic Environment*

The introduction of the abstraction levels also allows a more detailed discussion about different approaches predicting the perceived dynamic environment.

Current systems usually predict the dynamic environment mainly based on the estimated kinematic parameters of the tracked objects. This prediction works well for short-term horizons up to a few seconds and is part of the operational level within the proposed system architecture.

In more complex situations, which exceed simple lane keeping or a distance control system (e.g. turning in complex intersections), a more foresighted (up to

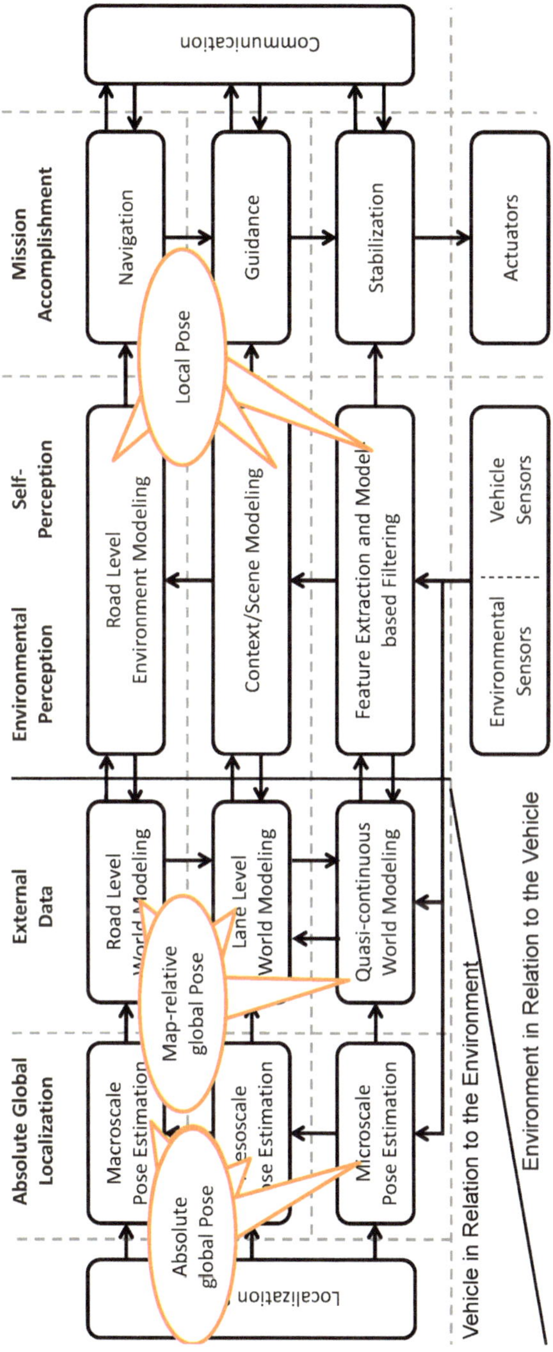

Fig. 5.3 Global localization solutions within the functional system architecture for an autonomous vehicle

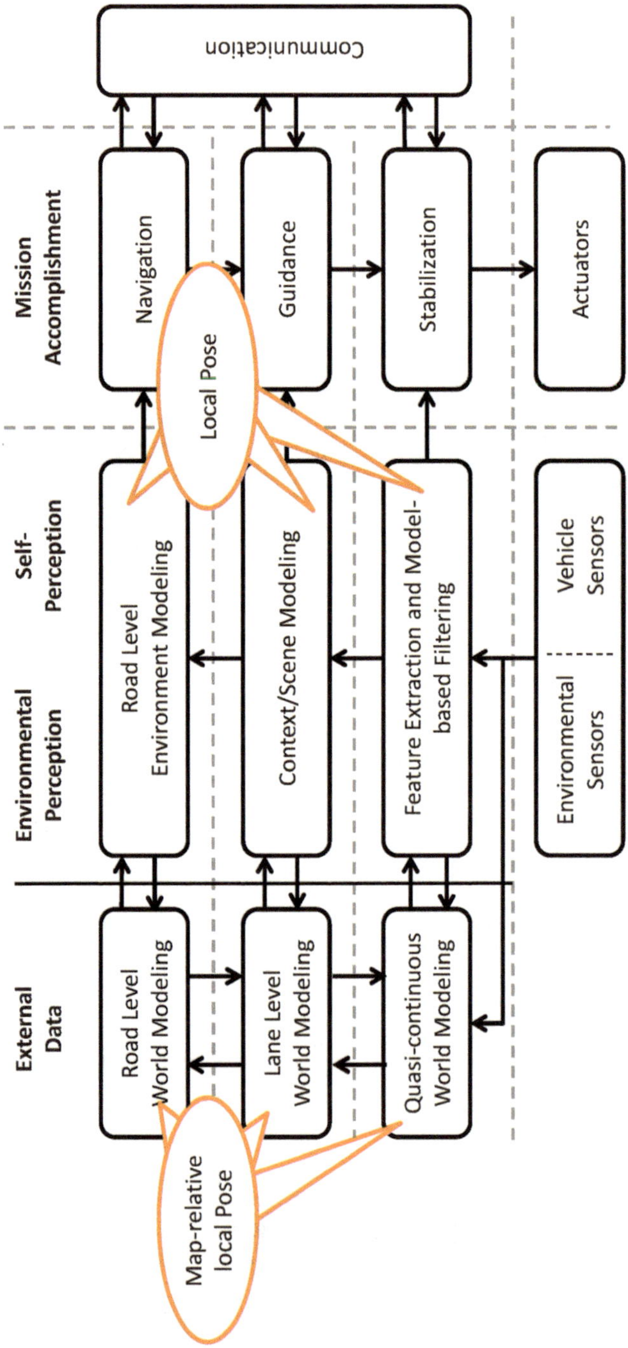

Fig. 5.4 Localization solutions within the functional system architecture for an autonomous vehicle without global localization

10 s) prediction is required, which is located on the tactical level of the system architecture. This medium term horizon is necessary to realize a comfortable and collision-free driving process, even at higher speeds (assuming that all traffic participants follow the traffic regulations). Within this time horizon up to about 10 s, the tactical decisions of the other traffic participants are the dominant parts (e.g. stopping at a traffic light, starting an overtaking maneuver). Furthermore, the currently estimated motion vector might not correspond to the real moving direction of the vehicle in the near future as the road course changes its direction. Assuming that the vehicle stays in its lane (maneuver decision or intention), the context information given by the course of the lane can enhance the prediction result as well. An example for such an approach is given by Herrmann and Schroven (2012).

Following this systematic (1-s prediction on the operational level and the 10-s prediction on the tactical level), a prediction on the strategic level might be possible, which predicts the dynamics of the vehicle's surroundings with a time horizon of several minutes or even longer. This would lead to a traffic-flow prediction on a road level (e.g. traffic jam predictions due to starting holidays or daily rush-hour traffic).

5.5.4 *Cooperation, Collaboration and Communication*

An elementary part of this functional system architecture is the systematic consideration of cooperative mechanisms independent of the technical realization (e.g. with explicit communication by optical or acoustic signals or V2X communication, or with implicit communication by gestures).

According to Matthaei et al. (2015) and Spieß (2014), "cooperation" describes a form of collaboration between at least two participating partners with the objective of finding a solution which is better referring to a previously defined and common goal. In public road traffic, cooperative behavior often includes some kind of trade-off between an optimal "global" solution for all partners and one's own intentions. Someone acts cooperatively if he or she does not insists on, but renounces his or her rights, for example, in a merging situation in dense traffic. In contrast to this form of cooperative acting or even interacting, we introduce the term "collaboration" for cooperative perception. However, both approaches (acting and perceiving) require some kind of communication.

The functional system architecture considers a bidirectional communication of the vehicle to its environment: it may communicate its intentions to other traffic participants (no matter if they are humans or machines) as well as to the passengers inside the vehicle (outgoing arrows of the column "mission accomplishment," see "communication" in Fig. 5.5, "explicit" communication) and it may also communicate its intentions by gestures (referred to as "implicit" communication in Fig. 5.5). Additionally, perceived environmental data on each abstraction level can be sent to a central service or to other traffic participants (outgoing arrows of

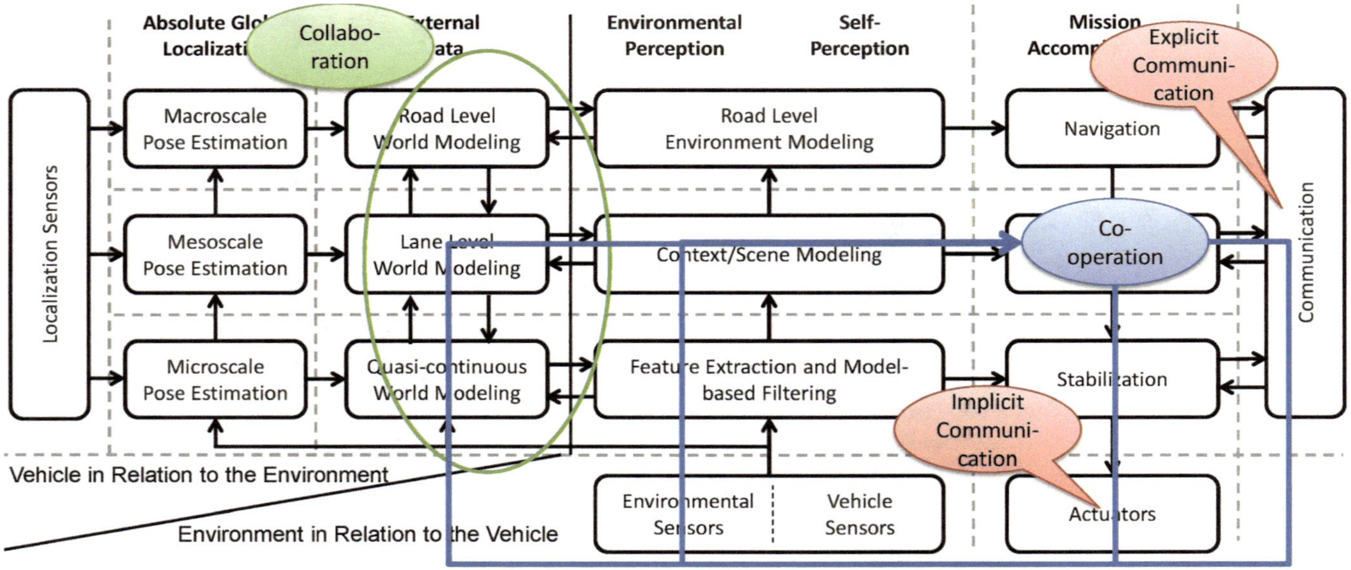

Fig. 5.5 Communication loops for co-operation (*blue*), collaboration (*green*) and communication (*red*) within the functional system architecture

the column "perception," see "collaboration" in Fig. 5.5). Sharing the perceived environmental information with other traffic participants (directly or indirectly via a central service) is the main idea of collaborative approaches (see e.g. Ko-PER[1]).

In addition to the output channels, the system architecture also provides appropriate input channels. On the one hand, information about the stationary and mobile environment can be received as external data and fused with the environmental data perceived locally. The intentions of other traffic participants are understood as a part of the external environmental data. On the other hand, instructions can be received through the communication interface to the mission accomplishment on each abstraction level. This might be the destination defined by the passenger on the strategic level, a lane change request by an emergency vehicle on the tactical level or a remote control on the operational level by an operator. Controlling the vehicle by an operator is not directly part of a cooperative behavior or autonomous driving, but it completes the description of communication-based interaction with the autonomous vehicle.

The activities, which are part of the commonly discussed cooperation, such as yielding right of way or letting others merge into the traffic flow, are mainly located on the tactical level in the module "guiding."

5.5.5 Self-Representation

The vehicle itself has to monitor its capabilities online for automated driving without supervision by a human being. The Chap. 6 by Reschka and Bagschik proposed a skill graph which can be interpreted as an abstracted functional model. Such a representation of the vehicle's current abilities is mainly located on the tactical level within the column "self-representation" in the functional system architecture and, thus, provides relevant data for the module "situation assessment" which is part of the block "guiding." For more information please refer to Chap. 6.

5.6 The Role of Rasmussen's Human Performance Model

5.6.1 Brief Introduction to the Concept of Rasmussen

Rasmussen introduced a model for the goal-oriented human behavior in the context of designing human-machine interfaces in Rasmussen (1983). His model is mainly a tool for predicting the human's performance and failures, but also helps to categorize certain tasks or solutions in the context of driving vehicles.

---

[1]http://ko-fas.de/deutsch/ko-per---kooperative-perzeption.html, 03/17/2015.

Rasmussen subdivided the human goal-driven behavior into three levels (Rasmussen 1983, p. 3):

1. On the *skill-based level*, the human acts "without conscious control". According to Rasmussen, the body acts "in most sensory-motor tasks [...] as a multivariable continuous control system synchronizing movements with the behavior of the environment" (Rasmussen 1983, p. 3). Feedforward control is mainly applied for rapid reaction, occasionally, feedback control is also necessary.
2. On the *rule-based level*, the human being behaves in well-known situations in a "goal-oriented" way. This means that one knows about a proven set of rules to solve certain situations. The rule-based level provides the input for the skill-based level.
3. In unknown situations, the human being no longer acts in a "goal-oriented" manner, but "goal-controlled", because proven rules are missing to solve the problem. In these cases, the human tries different approaches and checks the results against his or her goal. This procedure can either be carried out "physically by trial and error, or conceptually by means of understanding the functional properties of the environment and prediction of the effects of the plan considered" (Rasmussen 1983, p. 4).

5.6.2 Relevance Referring to the Driving Task

When looking at the levels of the driving task, they each have different timing constraints. An overview is given, for example, in Muigg (2009) based on the matrix of driving tasks published, for example, by Hale et al. (1990). According to Muigg (2009, p. 8), navigation might take minutes to hours, guidance tasks have to be executed within seconds to minutes and stabilization tasks within seconds.

Thus, the strategic navigation level provides such weak restrictions on timing that tasks can in principle be performed on the "knowledge-based" level. However, most searching algorithms, especially those for graph-search such as A* or Dijkstra, are typical implementations of rule-based approaches in the sense of Rasmussen. Given the road network, those algorithms generate the best route regarding certain constraints given by the driver (e.g. time, fuel-consumption, or distance). Once having found a solution for driving from A to B, the result can be stored in a look-up table. This would even simplify the route planning by using the look-up table next time.

The route to a desired destination provides a sequence of goals for the tactical level according to the driving task. At this level, tasks are typically executed on the rule-based level according to the performance level of Rasmussen. The consideration of the traffic regulations in well-known situations (e.g. while overtaking or stopping at a traffic light) is especially rule-based. However, there are some exceptions. One example is finding a path in a free space where lane-markings are missing (like the free-navigation areas of the DARPA urban challenge) which requires knowledge-based approaches. Another example is solving conflicts between traffic participants based on cooperative mechanisms or solving dilemma or polylemma situations (e.g. follow traffic regulations *or* avoid an accident), in which the ethical goal might be clear but the way how to achieve it has to be found out by "simulating" different possibilities. One can imagine that trying a lot of things (knowledge-based level) takes much more time than just applying a well-known rule (rule-based level).

Once having found the next target positions, a collision-free trajectory has to be generated. There are different ways of calculating such a collision-free trajectory. One is a direct lateral and longitudinal distance control, such as that performed by current adaptive cruise control or lane-keeping systems. This concept equals the way of acting on the skill-based level. Another way is to calculate a lot of possible trajectories to the next waypoint or to its close neighborhood and select the one which copes best with additional parameters such as comfort or safety. This approach would count more as knowledge-based acting, because the resulting trajectories are evaluated referring to additional goals (parameters). Examples are given by von Hundelshausen et al. (2008) and Werling et al. (2010).

Perhaps one can state the following concerning the computational effort of algorithms solving the same task: the simpler the approach itself, the more computational effort and time is required (brute force on the knowledge-based level). In other words: if performance-optimal implementations are required, more effort in developing algorithms is needed, so that the problem is no longer solved on the knowledge-based level, but on the rule-based or even skill-based level. Similar to the human's training for being able to solve tasks that are no longer knowledge-based, but rule- or skill-based, the developers of algorithms must somehow "train" the vehicle by implementing smarter algorithms.

One can also state something else: the human being needs to train due to its "computational" limitations, otherwise we would be too slow to survive. If we assume that machines might become much faster than humans, perhaps rule- or skill-based approaches lose their importance and most of the tasks are solved on a knowledge-base level (brute force). In this case, the machines would find an optimal solution and would probably compensate for the developers' laziness or human errors which might occur during the development of rule-based solutions for knowledge-based solvable tasks.

However, even today's computers show their limitations daily. Hence, current implementations depend on efficient algorithms which cope with limited computational power and, therefore, require high effort from their developers.

5.7 Advanced Driver Assistance Systems Within the Functional System Architecture

In this section we discuss how existing advanced driver assistant systems can be described with the functional system architecture in order to show the impact of the proposed functional system architecture. This discussion does not cope with the challenge of a self-supervising system, which is another task to solve when proceeding from assistant systems to highly or fully automated systems without human surveillance (see also Matthaei et al. 2015).

5.7.1 Navigation Systems

Current navigation systems support the driver on the strategic level (see Fig. 5.6). The driver can enter the destination desired via human machine interface. The task of the system is to calculate an optimal route based on predefined criteria (such as fastest route, shortest route or perhaps a trade-off) and to give the driver the corresponding navigation instructions (such as "turn left in 500 m") in a step by step manner. Today's navigation systems also provide an online adaptation of the planned route based on actual traffic flow information. The driver is responsible for conducting the vehicle. That means, the driver has to guide (tactical level) and to stabilize (operational level) the vehicle and, thus, remains responsible for collision-free driving according to local road traffic regulations.

The absolute global position is determined by a GNSS receiver with an accuracy of 10–20 m. The GNSS pose (position and orientation) is matched to the road-level map with the assumption that the vehicle stays on a road (so called "road-constraint"). This map-relative pose is relevant for the driver and the navigation system, not the absolute global pose acquired by the GNSS receiver. Some variants also use motion sensors (e.g. wheel speeds, acceleration sensors or gyroscopes) to enhance the global position or the matched map-relative pose based on the vehicle's trajectory.

Hence, navigation systems are mainly located on the upper level within the architecture. Only the motion estimation based on motion sensors is part of the operational level. However, the quasi-continuous data and lane-level information (e.g. lane-changes) of this estimation are irrelevant for the matching process into the road-level map-only prominent features, such as curves or turnings, can enhance the map-relative position as they describe the types of landmarks on the strategic level (see e.g. type one landmarks according to Hock 1994). That means, even though the motion estimation may be located on the operational level, the features used for the matching process are part of the strategic level and are implicitly extracted from the quasi-continuous trajectory representation during the matching process.

According to the view of the entire system of an autonomous vehicle, the integration of motion sensors into the localization processes for absolute and map-relative global localization are a kind of a shortcut, which is suitable for the development of a sub-system, such as this assistant system.

5.7.2 Adaptive Cruise Control

A common specification of an Adaptive Cruise Control (ACC) is described in ISO 15622 (2010). The task of the ACC system is to keep a certain distance (time-gap) from other traffic participants ahead. The system has to detect, locate and track other traffic participants in front of the vehicle and has to select the relevant target to be

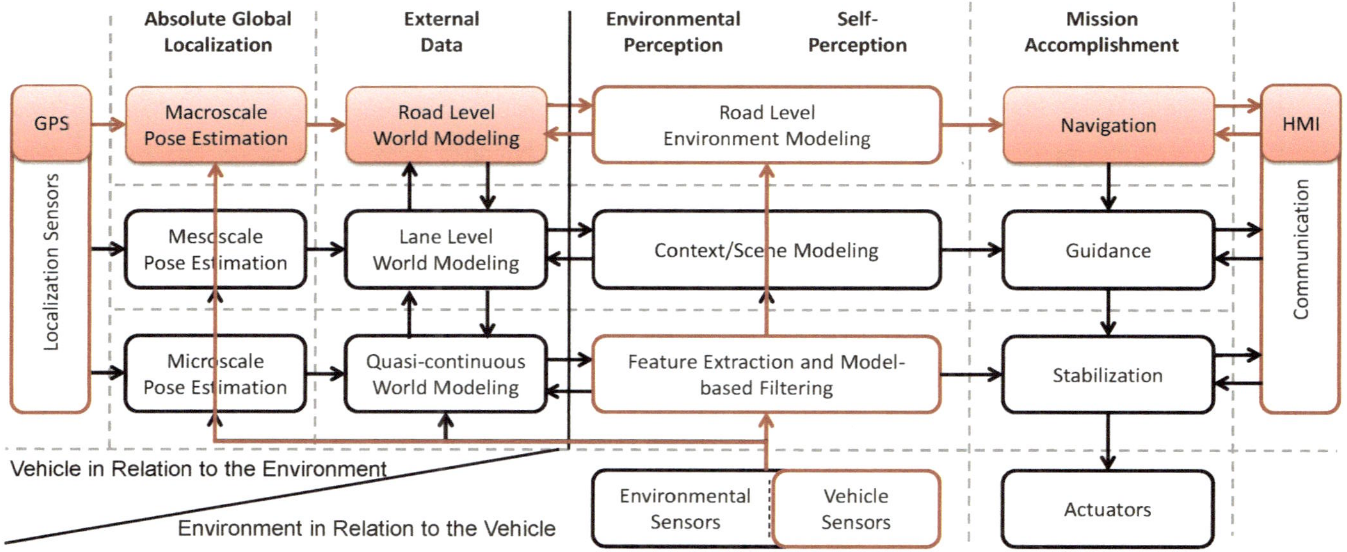

Fig. 5.6 Current navigation systems within the system architecture. *Red colored boxes* are part of the system, *red bounded boxes* are only partially covered by this system, and *black bounded boxes* are not part of the system

followed by the vehicle. The specification is explicitly restricted to highway scenarios, straight roads or curves with a constant radius. In these cases, the relevant target can be selected by a prediction of the host vehicle's position based on the current motion vector and the steering angle, because a constant radius is assumed. According to this specification, additional context information is not required. Stationary obstacles are not relevant for the system referring to the current specification (see also Winner and Schopper 2015).

The system is instructed by the driver on the tactical level. The driver defines the maneuver "following a target vehicle" by activating the system with a desired time-gap and a desired maximum velocity (see Fig. 5.7).

Even though many standard situations are handled by the system, the detection of traffic participants merging into one's own lane or selecting the correct target vehicle while changing the lane remain challenging tasks, as well as following the correct vehicle at the entrance of curves due to a changing radius.

It also becomes clear that this approach fails in many real-life situations. It is currently not specified that the system reacts according to local traffic regulations in all situations. Some example regulations, which are currently not covered by (most) ACC systems and not part of the ISO 15622 (2010), but relevant for a further automation of the longitudinal vehicle control, are given by the German traffic regulations:

- It is forbidden to accelerate while one is being overtaken (see §5, sub-paragraph 6 of the German road traffic regulations)
- It is forbidden to overtake on the right-hand side, except for (only some examples):
    - Inner-city streets
    - Near traffic lights
    - Overtaking streetcars
    - In traffic jams up to 60 km/h with a maximum speed difference of 20 km/h
    - On acceleration lanes (but *not* on deceleration lanes) on motorways
    - In the area of motorway junctions if there are wide lane markings and destination signs

All these regulations require a detailed knowledge of the context and they need a lane-based situation awareness and thus, are part of the tactical level. However, current ACC systems do not provide this context awareness. In other words, the driver has to ensure that the vehicle behaves according to the road traffic regulations, for example, by de-activating the system.

Only (reflecting) moving objects in front of the vehicle are detected and tracked on the operational level, and the context modeling is mainly done by matching the objects detected to the predicted trajectory of the host vehicle. This is used as the main indicator for the subsequent selection of the relevant object in the guidance block. In the stabilization block, the time gap to the target object is then controlled. The ACC does not, of course, provide any lateral stabilization.

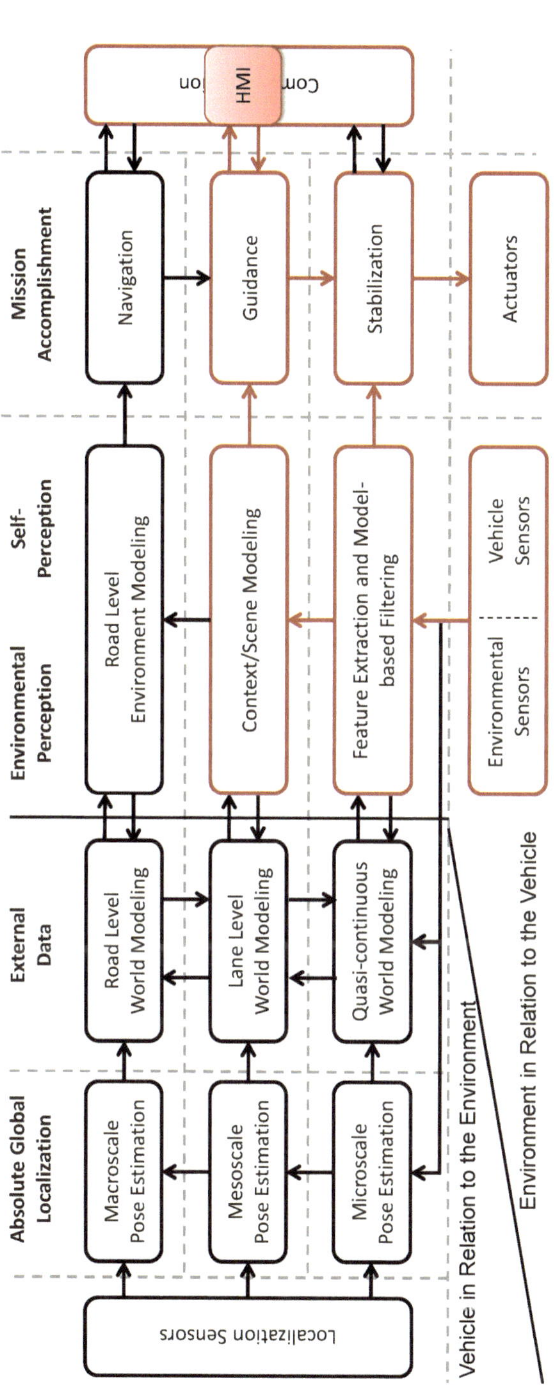

Fig. 5.7 ACC within the system architecture. *Red colored boxes* are part of the system, *red bounded boxes* are only partially covered by this system, and *black bounded boxes* are not part of the system. ACC which uses map data is not considered

The incompleteness of the ACC specification in comparison to the requirements for a system, which automatically fulfills the entire vehicle guidance and stabilization, is illustrated by marking the modules touched within the system architecture only with a colored boundary in Fig. 5.7.

Aspects of on-board diagnosis, such as surveillance of the sensors or the vehicle state, would be located within the column "self-perception." Depending on the system design, a detected malfunction of a system component might then lead to a degradation of the system performance or even deactivation of the system. The surveillance of the system boundaries is carried out within the block "guidance."

5.7.3 Lane-Keeping and Blind Spot Systems

Most lane-keeping systems are even less complex than ACC systems, as they generally do not touch the tactical level (despite onboard diagnosis). The selection of the lane which should be followed is decided by the driver. The system only stabilizes the vehicle within the lane selected.

Blind spot systems also do not include an explicit context modeling. According to ISO 17387 (2008) they estimate based on the driver inputs whether the host vehicle will move laterally soon and thus might collide with another vehicle next to the host vehicle or approaching from the rear. They thus mainly work in a "shared space," not explicitly considering the stationary environment (e.g. the exact course of the lanes as defined by lane markings). The corresponding relevant modules (except for the driver intention estimation which is not considered in this architecture for autonomous driving) are marked in Fig. 5.8.

However, first systems are available which combine these two systems. They also look for oncoming traffic or vehicles on adjacent lanes and "reduce the risk of unintentional lane changes"[2] and, thus, already touch the tactical level somehow.

5.7.4 Anti-Lock Braking System and Electronic Stability Control

Anti-lock braking (ABS) and electronic stability control (ESC) systems have been in series production for more than two decades. They directly support the stabilizing task within a subordinate control loop on the operational level. The environmental sensors are not relevant for these assistant systems, but these systems work based on motion sensors (see Fig. 5.9). For further details please refer to Zanten and Kost (2015).

---

[2]http://www.daimler.com/dccom/0-5-1210218-1-1210351-1-0-0-1210228-0-0-135-0-0-0-0-0-0-0-0.html, 03/22/2015.

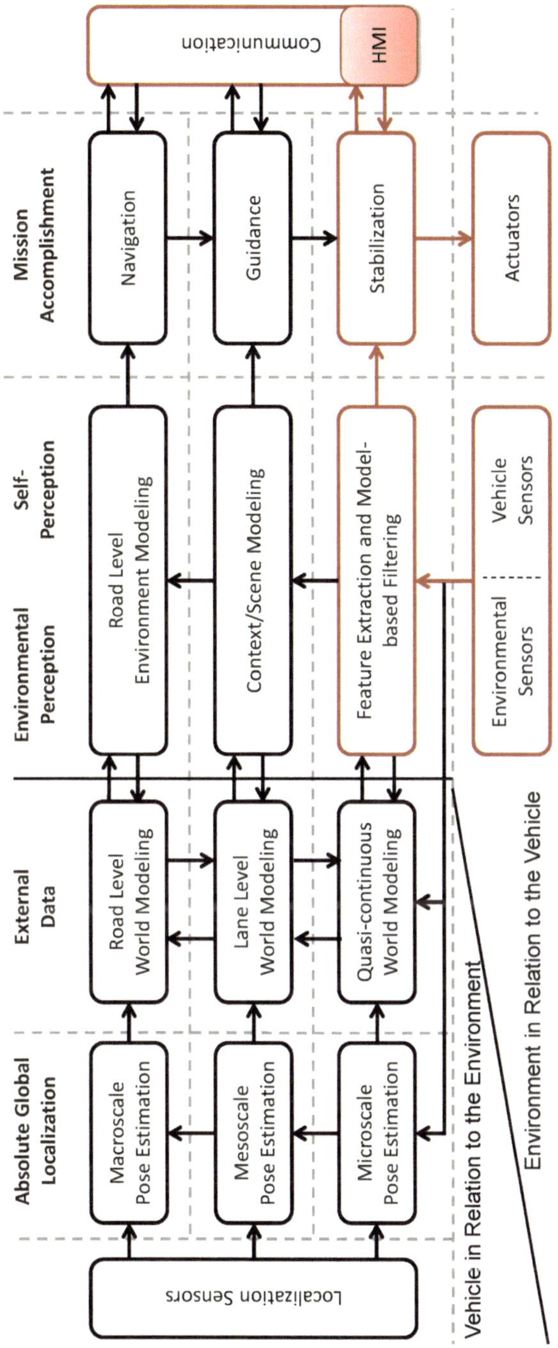

Fig. 5.8 Lane-keeping and blind spot systems in the functional system architecture marked in *red*

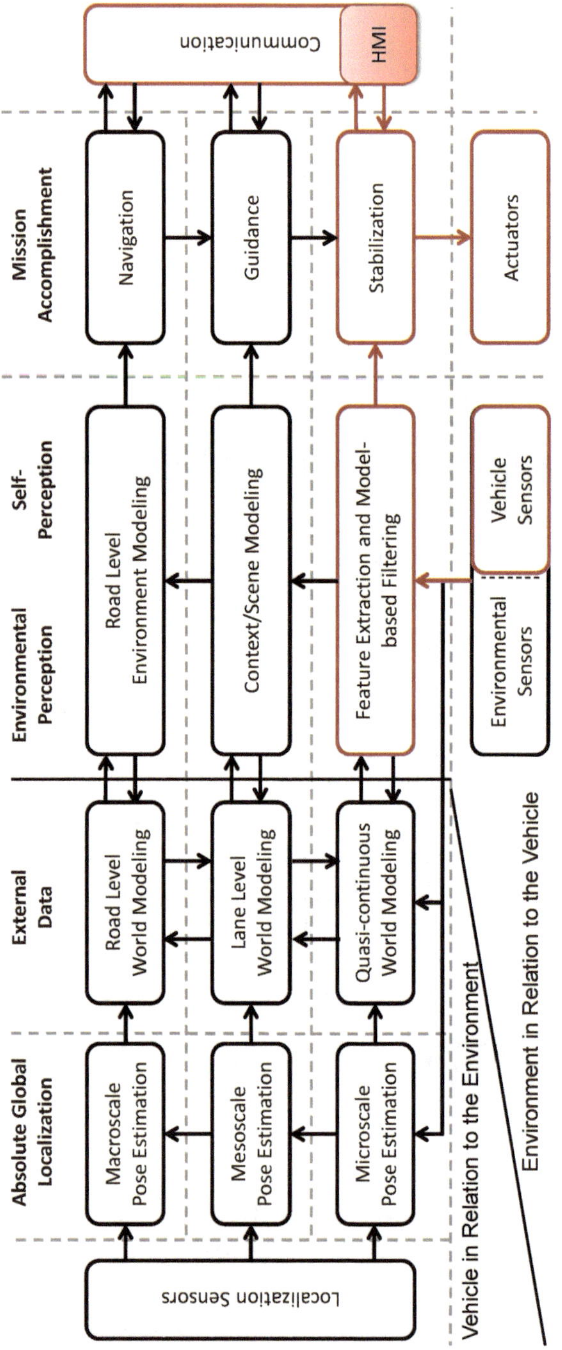

Fig. 5.9 ABS and ESC systems within the functional system architecture

Special challenges are provided by the estimation of the friction coefficient and the detection of the driver's intentions, especially for an ESC system. The system does not really know what the driver intends to do. It can only extrapolate the desired direction, based on the steering wheel angle and vehicle velocity.

5.8 Summary and Outlook

In this article, we proposed a holistic architecture for autonomous on-road motor vehicles which extends existing architectures by a systematic integration of external data, such as map data and V2X information. The consideration of a bidirectional communication also allows the implementation of automated map updates. Furthermore, we identified (as an extension to Moore et al. 2009) four different localization solutions which are covered by the proposed architecture.

The functional system architecture also points out two different ways of information flow for the driving task: one short loop directly from the environmental and vehicle sensors through the environmental perception to the mission accomplishment, and a second larger loop over a global localization and external data to the mission accomplishment. These two loops represent the different ways of a perception-driven or localization-driven approach for autonomous driving proposed in literature. We probably need both approaches for fully autonomous vehicles, depending on the safety requirements (e.g. if a fully redundant system is required or certain features can only be retrieved from map data).

The strategic level seems to be widely solved concerning the navigation systems and traffic reports, despite some enhancements concerning automated map updates, for example, and some issues regarding GNSS-based localization. The current research is mainly focused on special aspects within the operational level. Many details of the roughly drafted functional blocks are not yet solved from a technical point of view. In the context of highly automated systems according to level three of Gasser et al. (2012), time spans of about 10 s are discussed for a takeover by the driver. Based on this architecture, it becomes clear that a takeover due to an error on the operational level (e.g. because of a lane marking being suddenly missing, an unexpected obstacle or a pothole) is not possible. The takeover mechanism has to be designed within the tactical level.

The proposed system architecture also allows a further discussion about the role of the tactical level in future systems. Advanced driver assistant systems, such as an intersection assistant (see e.g. Herrmann 2013; Mages et al. 2015), make clear that larger time horizons are required for collision-free driving. The prediction only based on kinematic values does not yield usable results, especially in inner-city scenarios. In these cases, it is necessary to consider the scenery (e.g. lane course and traffic lights) at least or even the scene (including other traffic participants) for the prediction of the mobile environment. Additionally, the discussion about an ACC system, which works according to local road traffic regulations, demonstrates the increasing relevance of the tactical level, which also contains the consideration of the road traffic regulations and cooperative mechanisms. We thus expect future research activities to concentrate more on the tactical level, especially for advanced driver assistant systems and autonomous driving in urban environments.

References

Bacha, A., Bauman, C., Faruque, R., Fleming, M., Terwelp, C., Reinholtz, C., Hong, D., Wicks, A., Alberi, T., Anderson, D., Cacciola, S., Currier, P., Dalton, A., Farmer, J., Hurdus, J., Kimmel, S., King, P., Taylor, A., Van Covern, D., Webster, M.: Odin: Team VictorTango's entry in the DARPA urban challenge. J. Field Rob. **25**(8), 467–492 (2008)

Bonasso, P., Firby, J., Gat, E., Kortenkamp, D., Miller, D.P., Slack, M.G.: Experiences with an architecture for intelligent, reactive agents. J. Exp. Theor. Artif. Intell. **9**(2–3), 237–256 (1997). doi:10.1080/095281397147103

Broggi, A., Buzzoni, M., Debattisti, S., Grisleri, P., Laghi, M.C., Medici, P., Versari, P.: Extensive tests of autonomous driving technologies. IEEE Trans. Intell. Transp. Syst. **14**(3), 1403–1415 (2013). doi:10.1109/TITS.2013.2262331

Dickmanns, E.D.: Dynamic Vision for Perception and Control of Motion. Springer, London (2007)

Donges, E.: A conceptual framework for active safety in road traffic. Veh. Syst. Dyn. **32**(2–3), 113–128 (1999). doi:10.1076/vesd.32.2.113.2089

Du, J., Masters, J., Barth, M.: Lane-level positioning for in-vehicle navigation and automated vehicle location (AVL) systems. In: 7th International IEEE Conference on Intelligent Transportation Systems (ITSC), Washington, DC, pp. 35–40 (2004)

Gasser, T.M., Arzt, C., Ayoubi, M., Bartels, A., Bürkle, L., Eier, J., Flemisch, F., Häcker, D., Hesse, T., Huber, W., Lotz, C., Maurer, M., Ruth-Schumacher, S., Schwarz, J., Vogt, W.: Rechtsfolgen zunehmender Fahrzeugautomatisierung. Berichte der Bundesanstalt für Straßenwesen F83. Wirtschaftsverlag NW, Bergisch Gladbach (2012)

Hale, A.R., Stoop, J., Hommels, J.: Human error models as predictors of accident scenarios for designers in road transport systems. Ergonomics. **33**(10–11), 1377–1387 (1990). doi:10.1080/00140139008925339

Herrmann, S.: Kollisionswarnung im urbanen Straßenverkehr auf Basis einer probabilistischen Situationsanalyse. PhD Dissertation, Technische Universität Braunschweig (2013)

Herrmann, S., Schroven, F.: Situation analysis for driver assistance systems at urban intersections. In: IEEE International Conference on Vehicular Electronics and Safety (ICVES), Istanbul, pp. 151–156 (2012). doi: https://doi.org/10.1109/ICVES.2012.6294295

Hock, C.J.L.: Wissensbasierte Fahrzeugführung mit Landmarken für autonome Roboter. PhD Dissertation, Universität der Bundeswehr (1994)

ISO 17387: Intelligent Transport Systems – Lane Change Decision Aid Systems (LCDAS) – Performance Requirements and Test Procedures. Standard ISO 17387:2008. International Organization for Standardization, Geneva (2008)

ISO 15622: Intelligent Transport Systems – Adaptive Cruise Control Systems – Performance Requirements and Test Procedures. Standard ISO 15622:2010. International Organization for Standardization, Geneva (2010)

ISO 26262: Road Vehicles – Functional Safety – Part 4: Product Development at the System Level. Standard ISO 26262–4:2011(E). International Organization for Standardization, Geneva (2011)

Leonard, J., How, J., Teller, S., Berger, M., Campbell, S., Fiore, G., Fletcher, L., Frazzoli, E., Huang, A., Karaman, S.: A perception-driven autonomous urban vehicle. J. Field Rob. **25**(10), 727–774 (2008)

Levinson, J.S.: Automatic Laser Calibration, Mapping, and Localization for Autonomous Vehicles. Stanford University, Stanford, CA (2011)

Mages, M., Stoff, A., Klanner, F.: Intersection assistance. In: Winner, H., Hakuli, S., Lotz, F., Singer, C. (eds.) Handbook of Driver Assistance Systems: Basic Information, Components and Systems for Active Safety and Comfort. Springer International Publishing, Cham, Switzerland (2015)

Matthaei, R.: Wahrnehmungsgestützte Lokalisierung in fahrstreifengenauen Karten für Asssistenzsysteme und automatisches Fahren in urbaner Umgebung. PhD Dissertation, Technische Universität Braunschweig (2015)

Matthaei, R., Maurer, M.: Autonomous driving – a top-down-approach. Automatisierungstechnik. **63**(3), 155–167 (2015). doi:10.1515/auto-2014-1136

Matthaei, R., Reschka, A., Rieken, J., Dierkes, F., Ulbrich, S., Winkle, T., Maurer, M.: Autonomous driving. In: Winner, H., Hakuli, S., Lotz, F., Singer, C. (eds.) Handbook of Driver Assistance Systems: Basic Information, Components and Systems for Active Safety and Comfort. Springer International Publishing, Cham, Switzerland (2015)

Maurer, M.: Flexible Automatisierung von Straßenfahrzeugen mit Rechnersehen. PhD Dissertation, Universität der Bundeswehr München (2000)

Milford, M.J., Wyeth, G.F.: Mapping a suburb with a single camera using a biologically inspired SLAM system. IEEE Trans. Rob. **24**(5), 1038–1053 (2008). doi:10.1109/TRO.2008.2004520

Montemerlo, M., Becker, J., Bhat, S., Dahlkamp, H., Dolgov, D., Ettinger, S., Haehnel, D., et al.: Junior: the Stanford entry in the urban challenge. J. Field Rob. **25**(9), 569–597 (2008)

Moore, D.C., Huang, A.S., Walter, M., Olson, E., Fletcher, L., Leonard, J., Teller, S.: Simultaneous local and global state estimation for robotic navigation. In: IEEE International Conference on Robotics and Automation (ICRA), Kobe, pp. 3794–3799 (2009)

Muigg, A.: Implizites Workloadmanagement. PhD Dissertation, Technische Universität München (2009)

Müller, A., Himmelsbach, M., Lüttel, T., von Hundelshausen, F., Wünsche, H.-J.: GIS-based topological robot localization through LIDAR crossroad detection. In: 14th International IEEE Conference on Intelligent Transportation Systems (ITSC), Washington, DC, pp. 2001–2008 (2011)

Nothdurft, T., Hecker, P., Frankiewicz, T., Gacnik, J., Koster, F.: Reliable information aggregation and exchange for autonomous vehicles. In: Vehicular Technology Conference (VTC Fall), 2011 IEEE, San Francisco, pp. 1–5 (2011)

Rasmussen, J.: Skills, rules, and knowledge; signals, signs, and symbols, and other distinctions in human performance models. IEEE Trans. Syst. Man Cybern. **13**(3), 257–266 (1983)

Rauskolb, F.W., Berger, K., Lipski, C., Magnor, M., Cornelsen, K., Effertz, J., Form, T., et al.: Caroline: an autonomously driving vehicle for urban environments. J. Field Rob. **25**(9), 674–724 (2008). doi:10.1002/rob.20254

SAE International: Taxonomy and Definitions for Terms Related to on-Road Motor Vehicle Automated Driving Systems (2016)

Spieß, E.: Kooperation. Herausgegeben von Antonius Wirtz. Dorsch – Lexikon der Psychologie. Verlag Hans Huber. https://portal.hogrefe.com/dorsch/kooperation/ (2014)

Thrun, S., Burgard, W., Fox, D.: Probabilistic Robotics (Intelligent Robotics and Autonomous Agents Series). The MIT Press, Cambridge, MA (2005)

van Zanten, A., Kost, F.: Brake-based assistance functions. In: Winner, H., Hakuli, S., Lotz, F., Singer, C. (eds.) Handbook of Driver Assistance Systems: Basic Information, Components and Systems for Active Safety and Comfort. Springer International Publishing, Cham, Switzerland (2015)

Visintainer, F., Darin, M.: Final requirements and strategies for map feedback. Report D2.2. Ertico (2008)

Von Hundelshausen, F., Himmelsbach, M., Hecker, F., Mueller, A., Wuensche, H.-J.: Driving with tentacles: integral structures for sensing and motion. J. Field Rob. **25**(9), 640–673 (2008). doi:10.1002/rob.20256

Wachenfeld, W., Winner, H., Gerdes, C., Lenz, B., Maurer, M., Beiker, S.A., Fraedrich, E., Winkle, T.: Use cases des Autonomen Fahrens. In: Maurer, M., Gerdes, J.C., Lenz, B., Winner, H. (eds.) Autonomes Fahren – Technische, rechtliche und gesellschaftliche Aspekte. Springer-Verlag GmbH, Berlin (2015)

Werling, M., Ziegler, J., Kammel, S., Thrun, S.: Optimal trajectory generation for dynamic street scenarios in a Frenét frame. In: IEEE International Conference on Robotics and Automation (ICRA), pp. 987–993 (2010). doi:10.1109/ROBOT.2010.5509799

Wille, J.M., Saust, F., Maurer, M.: Stadtpilot: driving autonomously on braunschweig's inner ring road. In: IEEE Intelligent Vehicles Symposium (IV), San Diego, CA, pp. 506–511 (2010)

Winner, H., Schopper, M.: Adaptive cruise control. In: Winner, H., Hakuli, S., Lotz, F., Singer, C. (eds.) Handbook of Driver Assistance Systems: Basic Information, Components and Systems for Active Safety and Comfort. Springer International Publishing, Cham, Switzerland (2015)

Ziegler, J., Bender, P., Lategahn, H., Schreiber, M., Strauß, T., Stiller, C.: Kartengestütztes automatisiertes Fahren Fahren auf der Bertha-Benz-Route von Mannheim nach Pforzheim. In: 9. Workshop Fahrerassistenzsysteme, Walting (2014)

Part III
Functional Safety and Validation

Chapter 6
Towards a System-Wide Functional Safety Concept for Automated Road Vehicles

Andreas Reschka, Gerrit Bagschik, and Markus Maurer

Abstract In this chapter, a process to derive a system-wide functional safety concept for automated road vehicles is presented and a short introduction of Skill and Ability Graphs for a functional safety concept is given. The process to develop a functional safety concept contains an extension to the ISO 26262 standard's Driver Assistance System development process. This extension is a Skill Graph to model system skills in the concept phase. The Skill Graph improves the Hazard Analysis and Risk Assessment by modeling driving skills early in the development process. Additionally, the Skill Graph is transferred to an Ability Graph, used to design a self-perception and self-representation, which enables monitoring of the system's operation and functional capabilities online. This self-representation can be part of a technical safety concept. Based on the ability levels, safety actions can be derived which maintain or reach a safe state of operation. As a result, a self-monitoring system is possible, in which humans, either aboard the vehicle or external, do not have to monitor the system.

Keywords Automated driving • Functional safety • Systems engineering

6.1 Road Vehicle Automation

Road vehicle automation has been a research and development topic for many decades. The first ideas in the first half of the twentieth century could not be implemented due to technological restrictions (Mann 1958). Starting in the 1970s, camera technology found its way into vehicles and the first approaches in Europe, Japan, and the USA at universities and vehicle related companies showed up Dickmanns (2015), Tsugawa (1994). The PROMETHEUS project pushed the technology in the 1980s and 1990s and from then on the research and development activities spread widely across many universities and companies all over the world

A. Reschka (✉) • G. Bagschik • M. Maurer
Institut für Regelungstechnik, Technische Universität Braunschweig, Hans-Sommer-Str. 66, 38106 Braunschweig, Germany
e-mail: reschka@ifr.ing.tu-bs.de

H. Winner et al. (eds.), *Automotive Systems Engineering II*,
DOI 10.1007/978-3-319-61607-0_6

(Dickmanns 1994, 2002, 2007, 2015). Results of these activities were the first Advanced Driver Assistance Systems (ADAS) available in many vehicles today. Systems like Adaptive Cruise Control (ACC), Lane Keeping Support (LKS), and Forward Vehicle Collision Systems (FVCX) for warning, mitigation and avoidance of collisions use environment perception sensors. This separates ADAS from common Driver Assistance Systems (DAS) like Electronic Stability Control (Maurer 2012). The next step was the combination of such systems to increase the automation level of vehicles, e.g., in upper class vehicles like the Mercedes-Benz S-Class W222 (Schopper et al. 2013) and the Mercedes-Benz E-Class W213. These models feature the "DISTRONIC PLUS with steering assist", which takes over the longitudinal vehicle guidance and assists in the lateral guidance of the vehicle.

Until today, all ADAS rely on the human driver and his driving skills to control the vehicle immediately after the system deactivates itself, or to overrule the systems if they are not operating safely. As Ohl (2014, Chap. 2.1) points out, in all published research and development approaches to vehicle automation, a human safety driver is necessary to monitor the system permanently, either onboard the vehicle or from an external position.

To enable a safe operation, a holistic approach covering design phase, development, testing, and the operation of a vehicle guidance system for automated vehicles is necessary, starting with the definition of a use case, the possible scenarios in this use case, and the desired behavior of the vehicle in all operating scenarios. As this cannot be done in single components of the vehicle, the system as a whole has to be considered and interdependencies between functional components, hardware components, and software components have to be covered.

6.1.1 Definition of an Automated Road Vehicle

The technological development brought several terms for automated road vehicles, which should be clarified here shortly. A "vehicle guidance system" is an Electric/Electronic (EE) system, which takes over the whole driving task. The "level of automation" of vehicles is subject of publications from the German Federal Highway Administration (BASt), the National Highway Traffic Safety Administration (NHTSA) and the Society of Automotive Engineers (SAE) (Gasser et al. 2012; NHTSA 2013; SAE 2014). Bartels et al. (2015) summarize the terms in the draft of a glossary for the AdaptIVe project. Huang et al. (2007) and Huang (ed.) (2008) have published a more general approach to the definition of automation levels for unmanned systems in the "ALFUS" project at the National Institute of Standards and Technology (NIST) in the USA. Additionally, terms like "autonomous vehicle", "self-driving vehicle", "robot vehicle", and "driving robot" are used in this context and mean mostly the same.

The characteristics of an automated vehicle as understood in this chapter are the following:

- Motorized vehicle (e.g. car, bus, truck, motorbike)
- Not using rails or other mechanical or electrical guidance mechanisms
- No monitoring by humans necessary
- Not restricted to certain road type
- Not restricted to certain weather and road conditions
- Not demanding additional infrastructure, than infrastructure created for the human driver
- Operation in mixed traffic, with vehicles of all automation levels

In Matthaei and Maurer (2015), Matthaei (2015), Reschka (2016) and Matthaei et al. (2016) the functional requirements of automated vehicles with above mentioned characteristics are described in detail.

The automated vehicles covered in this chapter can be classified as SAE Level 4 or 5. Such vehicles can participate in all kinds of public traffic without a human driver or a teleoperator in normal operation and with the ability to maintain or reach a safe state for all involved humans from every situation. The functionality of such SAE Level 4 and 5 vehicles is described in Matthaei et al. (2016) and in the use cases in the Villa Ladenburg project (Wachenfeld et al. 2016), together with a summary of research projects in this field. Furthermore, the systems considered in this chapter cover all EE parts of an automated vehicle including actuators, sensors, communication interfaces, and controller hardware and software.

The systems covered in this chapter do not describe a retrofit solution, but combine the mechanical and electrical parts of a vehicle with the hardware and software necessary for vehicle guidance. It is assumed that the automated vehicle is developed as a singular holistic system. This holistic approach differs from the current series development, where singular Driver Assistance Systems are integrated into a vehicle and until today developed almost independently. In our understanding, the systemic approach is necessary due to the complexity of the vehicle guidance task and the experience resulting from the DARPA Urban Challenge 2007 (Rauskolb et al. 2008) and from the Stadtpilot project since 2008 (Wille et al. 2010; Nothdurft et al. 2011; Matthaei et al. 2016).

6.1.2 Definition of a Safe State

Maintaining and reaching a “safe state” during operation are the primary goals of a safety concept for automated vehicles. A safe state in the understanding of this chapter is a state where the operational risk is below a threshold which is accepted by society in the current scenario and the future development of the scenario. This level of acceptance is not yet defined and will not be in this chapter, but it is assumed that such a level exists.

The term "risk" is used as described in the ISO 26262 standard as "a combination of the probability of occurrence of harm and the severity of that harm" (ISO 26262 2011, Part 1). Grunwald (2016) points out, that the users of automated vehicles and other traffic participants experience the risks from automated vehicles in a passive role. From a manufacturer's point of view, risk has to be reasonable for customers and other stakeholders affected by the usage of the product. Thus, the reasonable risk of operation is not in the responsibility of users and other traffic participants, but of manufacturers.

There are three aspects of the risk of operation for the socio-technical system vehicle/passenger/traffic participants. On the one hand, the system has to operate safely in the current and future scenarios. For this operation, the internal system state and the current driving state regarding the traffic situation have to be considered. No passengers or other traffic participants may be endangered by the vehicle's actions. On the other hand, not yet visible, but possible, external events may lead to an unsafe driving state in the future, e.g., deciding a full stop of the vehicle could lead to a dangerous stopping place. Thus, the definition of a safe state has to consider the internal system state, the driving situation including the driving state, and the consequences of possible actions in the future. Additionally, a malfunctioning part/subsystem/element/function of the system can lead to unsafe behavior of the overall system. The functional safety concept has to prevent hazardous system states to maintain a safe state of operation.

Figure 6.1 illustrates a sequence of actions for the decision process. At start of operation, the system has to determine the current situation, its own capabilities, and events that can occur during operation in the current situation. Based on this information, possible actions can be determined. One of the actions is chosen and executed. If the action led to the final mission goal, operation may end. Otherwise, the process starts again.

The controllability of a vehicle by a driver or a teleoperator[1] according to the ISO 26262 standard is not considered, because automated vehicles in this context are not monitored by humans or a teleoperator permanently. Thus, the controllability of the automated vehicle by humans is not given. The controllability of situations for other traffic participants has to be considered as manually controlled vehicles can be affected by the actions of an automated vehicle.

Reschka and Maurer (2015) define the safe state further. Reschka (2016) describes an approach towards a safety concept for the use cases in the Villa Ladenburg project from Daimler-und-Benz-Stiftung, where the definition of a safe state is partly integrated. In this section the basic concepts are summarized.

If the risk of operation is below the reasonable risk, the vehicle is in a safe state of operation. In literature the term "risk-minimal state" is widely used. This term is confusing, because a minimization of risk would lead to actions reducing the functionality, like stopping the vehicle at a safe location. Figure 6.2 shows the

[1]A Teleoperator is a person, who monitors and even controls a vehicle via a wire-less communication link.

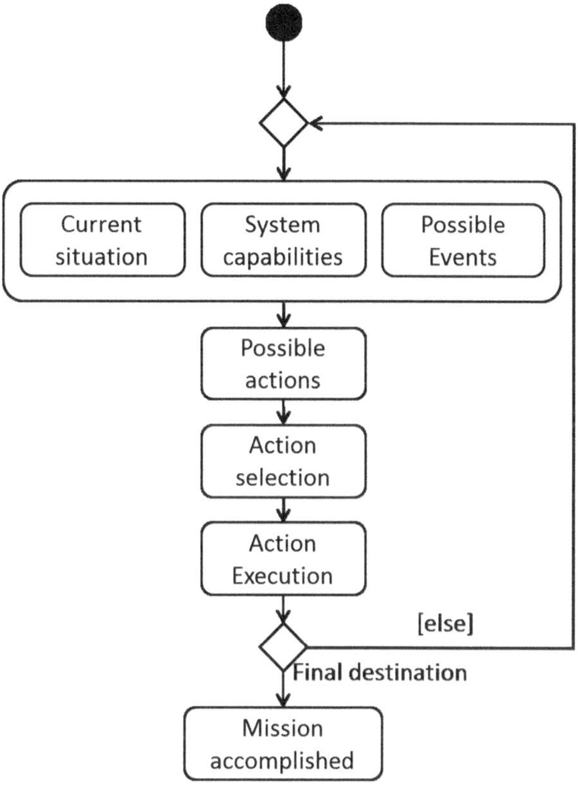

Fig. 6.1 Process of decision taking to maintain a safe state

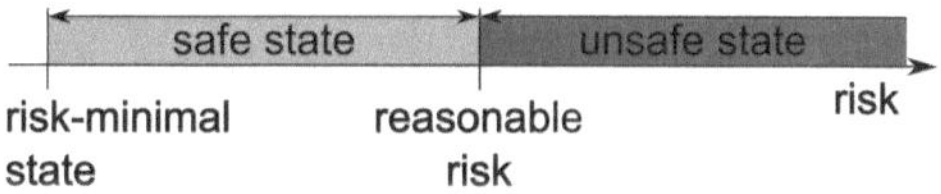

Fig. 6.2 Relation between risk, safe state, risk-minimal state and unsafe state

relation between a risk-minimal state and a safe state. Such a behavior is not desirable, because in this case the vehicle could not fulfill its mission, although slight changes in the driving strategy could result in an operation within a safe state with a reduced risk.

A main challenge of future research activities will be the definition of metrics, which represent the current operational risk and the current requirements considering the system state, the current situation, and the possible development of the situation based on possible actions of the automated vehicle. The question "How safe is safe enough?" has been asked, e.g., by Winkle (2016). Answering this question is still subject to further research.

An indicator for the current risk could be determined by comparing the situational functional requirements with the current functional capabilities of the vehicle. If the capabilities in the current driving situation are equal or higher than the requirements in the current driving situation, the vehicle is operated with a current risk, which is lower than the maximum reasonable risk and therefore within a safe state. Otherwise, the vehicle is in an unsafe state. This would lead to a situation-specific assessment of risk and a determination for an acceptable level must also be defined situation-wise. With increasing requirements and decreasing capabilities, the reasonable risk lies closer to the risk-minimal state. It is also possible, that the requirements are higher than the capabilities. This could lead to an achievable risk-minimal state, which is still unsafe. To estimate the current driving risk and the system boundaries a comprehensive self-perception and self-representation is necessary. Bergmiller (2014) summarizes this issue and the underlying technological challenges for x-by-wire vehicles. As automated vehicles are controlled by wire, the research results from Bergmiller are relevant for automated vehicles as well.

The generic set of conditions of a safe state derived from the use cases in the Villa Ladenburg project is presented in (Reschka 2016) and a further developed version is published in (Reschka and Maurer 2015). These are the following:

- The vehicle is controlled by a driver.
- The vehicle is controlled by teleoperation via Vehicle-to-Operator communication (V2O) (Matthaei 2014).
- The vehicle is driving automatically within its functional boundaries, especially with safe speed and adequate safety distances.
- The vehicle is stopped under the following conditions:
  - Relative speeds to other traffic participants are below a certain level.
  - The vehicle and its state are visible to other traffic participants.
  - The vehicle is not blocking emergency vehicles or emergency routes.

The conditions are generic and not connected to a special use case. Thus, for the analysis of use cases, for each scenario the safe state can be derived from the generic conditions. Smith (2016) also discusses the conditions necessary for a safe state, but from a legal perspective.

6.2 Process to Develop a Functional Safety Concept

Safety for automated vehicles has to be considered from two perspectives. Before the development process starts, the functional behavior of the vehicle in all possible scenarios in the functional range has to be defined. The primary goal of the resulting behavior is to avoid damage to persons and thus operate in a safe state. Other safety goals, like avoiding damage to property or following traffic rules, are secondary. This results in a value system, the vehicle has to follow, which also includes ethical considerations as described by Lin (2016). The resulting requirements define what the vehicle's necessary functional capabilities of the vehicle are. This process is not part of the ISO 26262 standard.

The ISO 26262 process starts with the assumption, that all functional requirements for the system under development are available and correct. By developing the system according to ISO 26262 functional safety for the desired behavior can be reached. The technical safety requirements define how the system has to be implemented to work properly. In this chapter, a process to develop a functional safety concept is presented. The identification of the desired behavior and the safe state for certain situations is not covered in this chapter, but subject to future research within the Stadtpilot project at the Technische Universität Braunschweig.

The development process according to ISO 26262 standard, which represents the state of the art for functional safety of EE systems, consists of different development phases. It starts with the concept phase including the Item[2] Definition, followed by a detailed design phase, the development of the item, test processes, and finally the operation of the item including the maintenance phase (ISO 26262 2011).

6.2.1 Product Lifecycle

The safety management lifecycle is covered in the ISO 26262 standard and provides a framework for the development process of Driver Assistance Systems (ISO 26262 2011, Part 2). The development process is based on the V-model, e.g., described in (VDI 2206 2004), and is applied in current series development in the automotive industry. The ISO 26262 standard includes methods to integrate safety into the development process. Although it is not yet clear whether the methods and tools described in the standard are applicable to automated vehicle systems with SAE Levels 4 and 5 as well, because there are no such systems in the market until today, the standard is used as a baseline for the following approach. In the proposed development process, multiple V-model based processes are combined. One V-model contains the overall development and product lifecycle. For the hardware development and the software development for each element[3] of the item another V-model process is used. As an automated vehicle is likely controlled by a distributed system, there can be several instances of the hardware and software V-model based development processes existing in parallel with interfaces described in the overall system development model.

The first main work product is the Item Definition (ISO 26262 2011, Part 3). It contains all relevant aspects of an item to be developed and defines the necessary behavior of the item during operation. After the item is defined, a Hazard Analysis and Risk Assessment follows, which results in the safety goals for elements of

[2]An item is a "System or array of systems to implement a function at the vehicle level, to which ISO 26262 is applied." (ISO 26262 2011, Part 1, 1.69).

[3]An element is a "System or part of a system including components, hardware, software, hardware parts, and software units." (ISO 26262 2011, Part 1, 1.32).

the item. The functional safety concept is designed to fulfill the identified safety goals. Thus, the safety goals have to be defined for the whole functional range of the item.

6.2.2 Defining the Scope of an Item

Before starting the Item Definition it is necessary to identify the item to be developed in the process of the ISO 26262 standard. For automated vehicles it is imaginable to determine the whole system as one item. Common practice in current development of ADAS is to divide the overall system into multiple subsystems and to develop a functional safety concept connecting those subsystems. The selected view on the whole system, either as one or multiple items, is fundamental for the following development process. Before the functionality of the item can be defined, it has to be clarified, if the whole system or subsystems are defined as items. According to the ISO 26262 standard an Item Definition shall contain complete specifications to other items, dependencies on and to other items and possible influences on the function of other items. By defining multiple items, the overhead due to multiple descriptions in each Item Definition would be difficult to handle. On the other hand, a complete system defined as one item has a very large Item Definition and more complex process steps for each development phase. Additionally, the layer which is used to define subsystems has to be chosen. It is possible to define subsystems on a functional level, a skill level, a hardware level, or on another level based on expert knowledge.

The ISO 26262 standard defines each system (not the Item) as a

> "*set of elements (1.32)that relates at least a sensor, a controller and an actuator with one another*" (ISO 26262 2011, Part 1).

The sensor(s) or actuator(s) can then only be a part of a system and not be systems themselves. This results in an item, consisting of several systems including perception, functional logic and actuators, which can basically sense, plan and act. It has to be mentioned, that, e.g., an environment perception system cannot be defined as an item according to the ISO 26262 standard, because it does not have an actuator and features only data acquisition and data processing. The lack of an actuator results in an impossible determination of an "Automotive Safety Integrity Level" (ASIL), because, if following the ISO 26262 standard strictly, no harm can be done by a system without an actuator. An ASIL is a requirement level used to classify safety goals, which result in different necessary measures to ensure the correct implementation of safety requirements for the system (ISO 26262 2011, Part 1).

The common practice in the automotive industry is to define the items based on the responsibilities of involved companies, e.g., an ACC system is divided into the functional EE items "ACC", "Electronic Stability Control", and "engine control" as all of them can be developed from different companies (Kriso et al. 2013). The

ACC item includes the environment perception and the functional logic. This results in an item ACC, which has no actuator in its scope, but is connected to the other items, which are actuators. This breaks with the definition of an item composed of sensor, controller, and actuator in the ISO 26262 standard, but is possible due to the option to define actuators as systems and items themselves. This possibility of being an item differentiates actuators from sensors in the scope of the ISO 26262 standard.

Another approach for the Item Definition could define all ACC relevant components, which can be parts of the systems Electronic Stability Control and engine control, in one singular item (Kriso et al. 2013). The result are different effects on hardware metrics and further process steps demanded by functions with high ASIL classification. Kriso et al. (2013) discusses this issue more detailed.

As the systems up to today are monitored by a driver permanently, the perception is assisted by the human perception skills and the functional logic is limited to adapting the vehicle speed to other traffic participants in front of the equipped vehicle. This results in lower ASIL for safety goals of the items ACC and engine control). The safety goals of the breaking system which ensure the ability to overrule the system for the driver and the absence of unintended braking result in high ASIL classification. The related hazards would be rated with high severity and low controllability for the driver and other traffic participants.

For an automated vehicle, where no driver is needed, the reduction of the ASIL for the functional logic including the perception and the engine control is not applicable. Thus, either defining the whole system as an item or dividing the system into multiple items *seems* to result in an ASIL D classification. Additionally, actuators like the wiper, the windshield washer system, the lighting, the brake lights, the indicator, the horn and others have to be considered as well, because they have to be controlled by the vehicle guidance system.

In case of the DISTRONIC PLUS with steering assist from Mercedes-Benz, technically the system is divided into an ACC and a LKS. The ACC could be developed as described above. The LKS features a safety critical steering system to counteract departure of the current lane. As the driver monitors the system and the system only assists and does not take over the whole steering task, again the perception, functional logic, and motor control is not as safety critical as in a system which does not have to be monitored permanently. Nevertheless, the steering torque must be controllable for the driver at any time, which results in a higher ASIL classification for the limitation function. If using these subsystems for an automated vehicle it is probable, that they have higher safety requirements, because the controllability by a driver is not applicable.

To sum this up, one approach to determine ASIL to functional requirements could be to define a vehicle guidance system as one item and its components as elements. This holistic approach results in an Item Definition, which has to cover all aspects of the vehicle guidance system. The follow-up processes will treat the item in the same manner and so the Hazard Analysis and Risk Assessment and the functional safety concept are for the whole vehicle guidance system. A main benefit of this approach is to identify risks and harms for the whole system. A main deficit is the complexity and the fact that for a complex system functionality an open set of situations has to be considered. It is quite easy to find situations, which require reliable functionality determining ASIL D (highest). Assumed that every necessary functional component can be realized in ASIL D somehow, functional deficits result in a reduced usability compared to the possibilities of human drivers.

Figure 6.3 illustrates a left turn situation from the city ring road in Braunschweig, Germany. The ego vehicle is approaching from the top left corner (Cyriaksring) in the green lane. At the red area in front of the tram rails vehicles turning left into Luisenstraße typically stop. If an automated vehicle would have to turn left, the vehicle could have to stop there as well, because approaching vehicles from the bottom right in the opposite direction of Cyriaksring could approach. As illustrated in Fig. 6.3, the distance from the stopping point to the stop line for the opposite direction is about 44 m. Although exceeding the speed limit is not allowed by traffic rules, it is common in this situation that vehicles approaching drive up to 20 m/s. Thus, the ego vehicle would have only about 2.2 s to cross the opposite lanes and drive into Luisenstraße. As there is a pedestrian crosswalk and a bicycle path, both depicted as orange area in Luisenstraße, the ego vehicle could have to stop at the second red area. As a result the vehicle guidance system in the ego vehicle faces high functional requirements regarding viewing distance and reliable dynamic object tracking. Especially important within the context of this chapter,

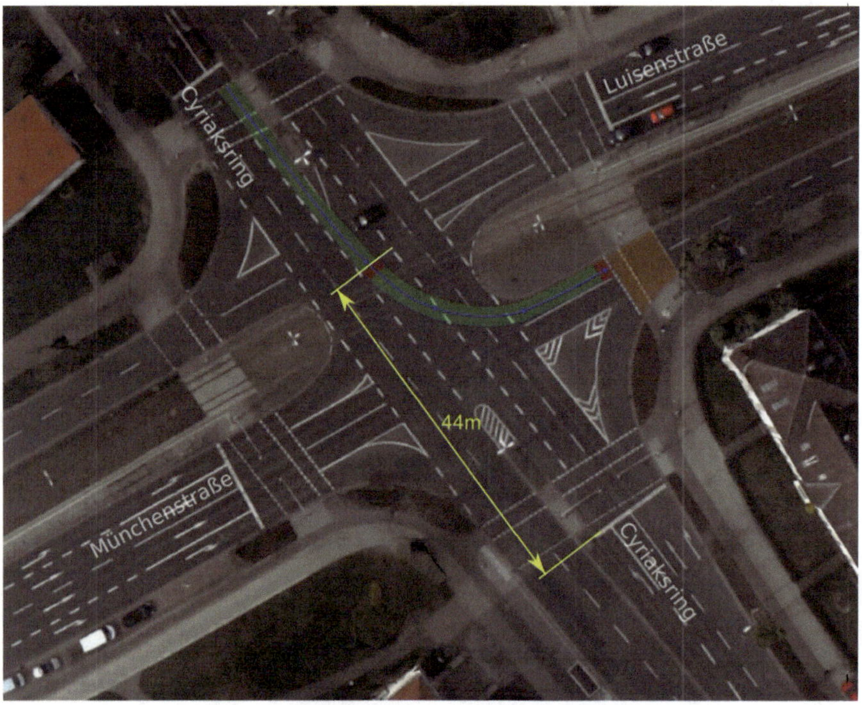

Fig. 6.3 Exemplary situation, Geographic data © 2015 GeoBasis-DE/BKG (@2009), Google

high reliability requirements for the whole system arise, because a failure or a misinterpretation in this situation could cause a severe accident. The illustrated situation is not artificial and can likely be encountered in almost every drive with the automated vehicle. In our understanding of safety requirements, this would result in an ASIL D for all of the components of the vehicle guidance system involved in solving this situation safely.

6.2.3 *Work Products in the Concept Phase*

Figure 6.4 shows input, process steps, and work products in the concept phase of the proposed development process. As a new element in comparison to the ISO 26262 standard the Skill Graph is added to the work product list. Green boxes are constraints from the product idea. Yellow boxes show work products. Blue boxes show process steps of the ISO 26262 standard. Solid lines show unlimited usability for a follow-up product, dashed lines show limited usability.

In general the development process for Driver Assistance Systems proposed in the ISO 26262 standard can be applied to automated vehicle development as well, but there are several facts about automated driving which need special consideration:

Open System

A vehicle guidance system for an automated vehicle is likely going to experience an open set of scenarios in public traffic. Starting with the Item Definition, a complete as possible definition of the expected operation environment is necessary to cover a large number of scenarios. Besides this, identifying the relevant parameters can be helpful.

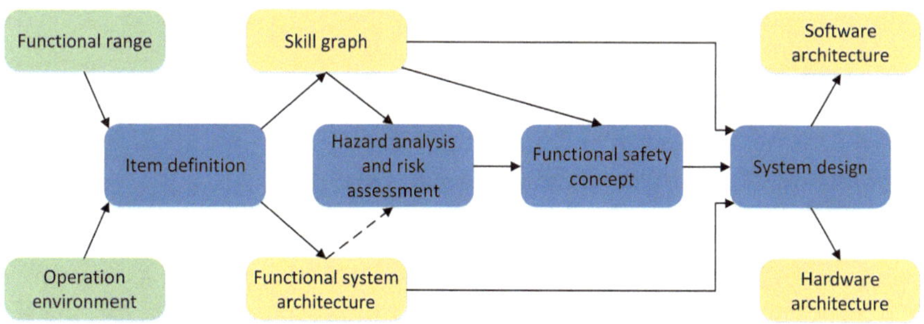

Fig. 6.4 Work products and process steps in the development process; *Green Boxes* are input to the development process, *blue boxes* are process steps, *yellow boxes* are work products; *arrows* show the sequence of actions and the usage of work products and data

Legal Issues
For an universally usable driverless vehicle the operation should be possible on every road in an existing law-enforcement area, e.g. all roads in Germany. For other countries, their respective laws and regulations have to be implemented as well, to create a framework for operation. Furthermore, changes to the legal issues during time of operation must also be considered and an update process is necessary.

Road Network
Besides regulations and laws the static structure of the road network, the traffic control infrastructure, and the roadside structures have to be considered.

Environment
Environmental conditions like sunlight, condensation, and temperature have to be considered due to their influence on perception.

Road Conditions
The road surface and dirt can have an influence on the friction coefficient and the perception of road markings.

Traffic Participants
The current and possible behavior of other traffic participants has to be considered.

Controllability
The controllability of the vehicle in emergency situations by a human driver is not given. Thus, only the controllability of the situation by other traffic participants can be considered. Recommended safety analysis methods like a Fault Tree Analysis or a Failure Mode and Effect Analysis have to cope with a complex system. Their application may be a source of development errors due to an imperfect consideration of above mentioned parameters.

Distributed Development
As sensors, actuators, electronic control units, mechanical parts and software are developed by different companies, system integration and monitoring is difficult and the development effort has to be coordinated.

6.2.4 Safety in the Concept Phase of an Item

According to the ISO 26262 standard, the concept phase contains the Item Definition, which describes the operation environment and the functional range of the developed item. Additionally, interfaces to other items in the overall system have to be defined. After the Item Definition, a Hazard Analysis and Risk Assessment is conducted to analyze the item. The result is a collection of operation risks for the developed item and a determination of ASIL for the elements of the item.

Functional System Description

In the concept phase, the external behavior of the item, possible operating modes, and the operation environment is analyzed and described. For a complete description a textual part and a Functional System Architecture are common. In this chapter, the Skill Graph is added to the functional description.

Textual Description

The textual functional system description is the first work product in the Item Definition according to ISO 26262. It contains the functional range of the item, its operating modes and a description of the operation environment. This includes the system boundaries, either functional or resulting from the operation environment.

The operation environment contains a description of the road network, the traffic infrastructure and the traffic participants relevant for the desired functional range. Besides these, regulations, rules, laws and standards that apply are at least referenced. Besides written rules, a value system as proposed by Gerdes and Thornton (2016) is necessary, which has to be considered in driving decisions. The goal must be to describe the system and its environment as complete as possible.

For an automated vehicle for public traffic operation, a complete set of the possible situations or a meta-set of possible elements and their characteristics is beneficial. Dickmanns (2007) gives different lists for road types, traffic participant types, road conditions, and weather conditions. The possible combinations are so manifold, that it is unlikely to describe all of them. To avoid requiring all combinations, safety can be enabled with mastering of so called pathological scenarios. These are scenarios that occur rarely, but can have severe consequences and thus have to be considered. A methodology to find such pathological scenarios for the item description could be to use accident data (Winkle 2016) and realistic driving studies like EUROFOT (Benmimoun et al. 2012). Utilizing this data only covers recorded accidents and dangerous situations encountered by human drivers. For a vehicle guidance system, additional dangerous situations are imaginable, which are not yet identified. An approach to a structured derivation of situations was presented by Schuldt et al. (2014) and in Chap. 7 in this book.

Another aspect is the definition of a safe state for each of the possible situations. If normal operation is no longer possible, there have to be actions and possible transitions to keep the vehicle in a safe state.

Skill Graph

For the self-perception, Reschka et al. (2015)[4] describe a concept based on Skill and Ability Graphs. They show how to describe skills of the resulting system and the possibility to determine an ASIL for each of the skills instead of the whole system.

[4]Reschka et al. (2015) used the terms skill and ability interchanged than this chapter. Due to new research results, the terms had to be switched. Skills are defined in the concept phase and abilities are used during operation of the vehicle.

The Ability Graph has been introduced by Pellkofer (2003) and Siedersberger (2003) for automated vehicle guidance and was applied to a full drive-by-wire-vehicle and further improved by Bergmiller (2014). By Reschka et al. (2015) the skill and ability concepts were utilized to fit to our understanding of automated vehicles. From the resulting description of the automated vehicle system, the Skill Graph can be derived.

Pellkofer (2003) introduced the graph as an ability network (German: Fähigkeitennetz) as a part of the ability concept (German: Fähigkeitenkonzept) developed by the working group "Verhalten" (English: Behavior) at the Universität der Bundeswehr München (Siedersberger et al. 2000). The whole concept was introduced by Siedersberger (2003) and uses abilities from the categories perception and scene description, behavior decision, driving and saccadic vision, and planning as introduced by Maurer (2000) to model the abilities of an automated vehicle during operation. Bergmiller (2014) further developed the concept to a self-concept, which enables a detailed self-representation of a full-by-wire actuation system. Bergmiller (2014) calls the abilities of Pellkofer and Siedersberger *skills* and divides them into three groups: basic skills, action skills, and behavioral skills. These groups represent the three levels of performance introduced by Rasmussen (1983).

A starting point for abilities are the necessary driving maneuvers for the desired functionality and the relevant parameters to each of the identified driving maneuvers. As Dickmanns (2007, p. 43) proposes, a driving maneuver is a specific control scheme in the system and each time, the control scheme changes, another maneuver is executed. Nagel and Enkelmann (1991) defined a necessary set of 17 driving maneuvers for vehicles in public traffic. Tölle (1996) defined a basic set of nine driving maneuvers derived from Nagel and Enkelmann (1991). For the definition of the Skill Graph, basic driving maneuvers according to Tölle (1996) can be used for the top layer in the Skill Graph. For each of the basic maneuvers, the necessary underlying skills can be derived from human tasks executed to fulfill the top skills. These skills and their properties describe the system more detailed and in a structure, which can be transferred to a hardware and software architecture and can be utilized for online self-perception and self-representation of the vehicle.

Before introducing an exemplary Skill Graph for a Driver Assistance System, the difference between a functional component and a skill should be clarified. A functional component provides a functionality within a system. Many functional components work together to combine their functionalities to a complex system performing more complex tasks than a single functional component or a subset of functional components. To give a systematic overview of a complex system, the Functional System Architecture can be used as a basis, but it is not easily possible to identify safety relevant information like components with single points of failure or other metrics for a safety analysis, because the system architecture does not model the performance dependencies between functional components necessary for a safety analysis.

A Skill Graph provides this information as it contains performance metrics and in the graph. Redundancies can be identified, because each skill relies on at least two other skills and is itself necessary for more complex skills. Additionally, the graph, which is built by deriving necessary skills for the item in development allows an identification of cyclic dependencies, e.g., if a skill is necessary for another skill and relies on this skill itself. Especially considering self-perception these cyclic dependencies are important to identify. A detailed analysis of the dependencies between skills is possible as well. These dependencies show which node in the tree can be removed without causing a complete outage of the system. It is possible to identify the most important nodes and the existence and the demand of redundant paths in the graph. Thus, an ASIL determination for skills is possible and even a skill-based ASIL decomposition seems feasible.

For every skill one or more aggregated performance metrics are necessary, which represent the current capability of a skill. These performance metrics use performance values from other skills and additional sensor values.

Figure 6.5 shows a Skill Graph for an Adaptive Cruise Control system, which includes Cruise Control and Electronic Stability Control functionality. The ACC can follow another vehicle in the estimated path of the ego vehicle based on the current speed of the leading vehicle and the yaw rate of the ego vehicle. Green blocks are meant to be the most abstract skills or the maneuver. Blue blocks describe skills with performance metrics, which are necessary to fulfill the top-level skill. Yellow blocks describe actuators and orange blocks sensors or inputs to the system.

The top node is the skill ACC driving. This skill enables the driving maneuver "Follow" (German: "Folgen" according to Tölle (1996)) and additionally enables Cruise Control and Electronic Stability Control.

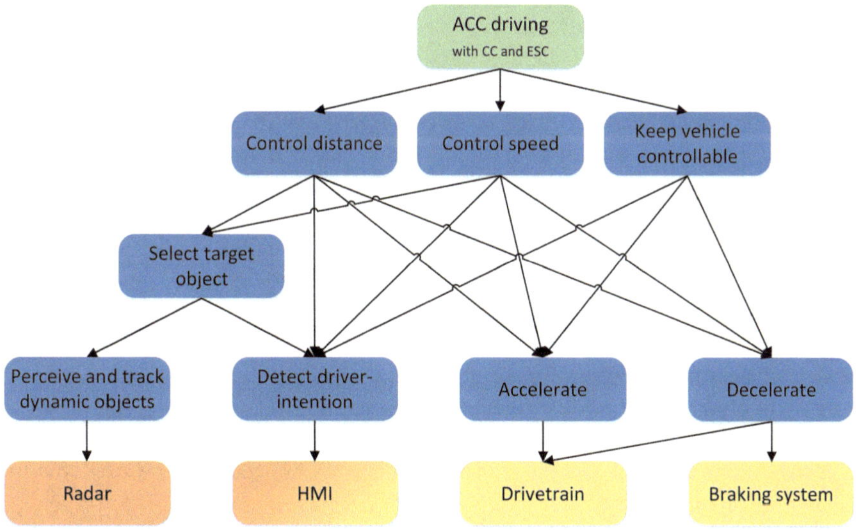

Fig. 6.5 ACC Skill Graph, which includes Cruise Control (CC) and Electronic Stability Control (ESC) functionality

A Cruise Control functionality (skill: "Control speed") is active, if no leading vehicle is available.

An Electronic Stability Control functionality (skill: "Keep vehicle controllable") is permanently available and acts if the vehicle is about to get out of control.

"Control distance" is the skill to adapt the ego vehicle's speed to a leading vehicle, which has to be selected (skill: "Select target object"). Additionally, the skills "Accelerate" and "Decelerate" are necessary, and the driver is able to define a maximum speed and the time gap to a leading vehicle (skill: "Detect driver intention"). "Control speed" is the skill to drive a desired speed if no leading vehicle is available. Thus, the skill depends on the same skills as "Control distance". "Keep vehicle controllable" is the skill, which limits vehicle dynamics in a way that the driver is able to control the vehicle in every situation immediately. This skill depends in the ACC system only on the longitudinal vehicle guidance ("Accelerate", "Decelerate"). In Fig. 6.5 the skill "Select target object" is exemplary composed by "Perceive and track dynamic objects" and "Detect driver intention". The perception of objects is necessary to identify relevant objects, the driver intention is necessary to detect the desired path of the driver, because then it is possible to select the relevant target object. An early version of a similar approach has been published in 2012 (Reschka et al. 2012a, b). As mentioned before, our current understanding of Skill and Ability Graphs is described in Reschka et al. (2015).

Functional System Architecture

Although not required explicitly in the standard, it is helpful in the concept phase, to design a Functional System Architecture. The ISO 26262 standard foresees a system architecture in the design phase. From the description in the standard, it is not clear which kind of architecture is used for the overall functional system modeling. For an automated vehicle such a system architecture based on the work by Maurer (2000), Wille (2012), Matthaei (2015), Matthaei and Maurer (2015), and others is described in Chap. 5 of this book. The Functional System Architecture lacks one important aspect for safety processes. There is no explicit representation of safety components as these are integrated into the functional modules (Matthaei 2014). This results in a more difficult identification of safety critical parts, because it is not visible which consequences failures have in the components of the Functional System Architecture. Therefore, it seems advantageous to use a Skill Graph as a modeling tool, which compensates this deficit. The Skill Graph can be understood as a link from the functional requirements in the Item Definition to the design phase of the system. Due to the fact, that a common understanding of the parties involved in the development process is necessary, a system architecture can help to build this. In the remaining of this chapter it is assumed that a Functional System Architecture is available in the concept phase and can be used for the following development phases.

6.2.5 Hazard Analysis and Risk Assessment

The Hazard Analysis and Risk Assessment, as the next step in the ISO 26262 standard development process, can be fulfilled based on the Functional System Architecture and the Skill Graph. The first step is to identify hazardous events during operation of the item. Due to modeling of the skills in a graph each skill can be evaluated in relation to the whole system. The process of the Hazard Analysis and Risk Assessment for automated vehicles with SAE Level 4 or 5 is subject to our future work. Bagschik et al. (2016) describe an approach, which uses Skill Graphs to identify malfunctions of the system based on the Item Definition. They combine these malfunctions with possible operating situations to identify hazardous events.

6.3 Ability Graphs as Part of a Functional Safety Concept

Integrating functional safety into a system is possible by adding hardware and software components which influence the functional operation little. These safety functions should detect and eliminate failures and keep the operational range as high as possible. To detect degraded performance caused by internal and external events, a self-perception is necessary. To determine the impact of those a self-representation is necessary. This can be done based on skills and abilities. To restore performance, reconfiguration and other self-healing methods can be used.

6.3.1 Related Work

A system-wide safety concept for SAE Level 4 and 5 automation is not known to the authors and thus, it is only possible to discuss publications which cover parts of such a concept. For automated vehicles (Hörwick and Siedersberger 2010) presented a safety concept for parts of the system and without considering the development process. This concept is one of the first for automated vehicles with SAE Level 3 automation and is applied to a traffic jam assist function which controls the vehicle in traffic jams up to a speed of 60 km/h (Hörwick 2011). The system is capable of reducing the driving risk by stopping the vehicle on the current lane with the usage of action plans. As the system is used in traffic jams only, the automated vehicle is likely followed by others in a traffic jam and therefore high relative speeds to others are not very probable. A main difference to driverless vehicles unsupervised by humans is the availability of a human driver, if the system reaches its functional system boundaries. In the proposed safety concept several monitoring instances are integrated into the EE system which monitor the operation of system components. However, the concept is not directly applicable to driverless vehicles, because of a missing fallback solution if the traffic jams ends. Furthermore, the proposed concept is a technical safety concept without considering the development process of the EE system and a methodology for identifying functional requirements to identify critical situations.

The research and development project aFAS (German project title: Automatisches fahrerlos fahrendes Absicherungsfahrzeug für Arbeitsstellen auf Autobahnen, English translation: Automated driverless protective vehicle for motorway hard shoulder road works) is one of the first projects which aims at a real world usage of an unmanned vehicle in public traffic (Ohl et al. 2012; Stolte et al. 2015a, b). Although the use case is limited to the hard shoulder of motorways, the unmanned operation without permanent monitoring is new to the automotive industry. In the project, a study on the applicability of the ISO 26262 standard to fully automated driverless vehicles is one of the research goals. As the prototype is not yet operated unmanned, no real-world results are available. The resulting system will enable SAE Level 4 automation and the safety concept is subject to our future work.

In Maurer (2012) and Reschka et al. (2015) a development process for collision avoidance and mitigation systems is described, which is applied in current development of the automotive industry. The chapter focuses on a development process, which is driven by the customer needs, technologically feasible, and economic constraints. Due to the technological difficulties that can arise in the development process of Advanced Driver Assistance Systems, an iterative approach could possibly avoid expensive restarts of development phases. This iterative approach will likely be necessary in the aFAS project as well and also for our future work in the Stadtpilot project.

A first approach based on the ISO 26262 standard has been presented in Reschka et al. (2011) for the Stadtpilot project. This approach was used to develop and test a prototype automated vehicle for operation in public traffic. The main focus was the approval process of the control software. In Reschka et al. (2012a, b) a monitoring concept has been introduced in the Stadtpilot project, which allowed basic system monitoring based on heartbeats, timing monitoring, and the calculation of aggregated performance criteria for certain system functions. This concept has to be improved further to deal with all aspects of higher levels of driving automation, especially concerning the absence of a human driver.

6.3.2 Utilizing Ability Graphs to Improve Safety of Operation

During operation of an automated road vehicle, a self-representation of the vehicle is necessary to improve safety of driving decisions. This self-representation is implemented with the Ability Graph, which is derived from the Skill Graph. The current performance of each ability is measured with a performance metric. In combination with the monitoring of hard- and software components these performance metrics are collected in the self-representation. In addition, skill specific metrics are passed to depending skills. With this concept, complex abilities can identify functional deficits and thus are able to consider this reduced performance level in decision tasks. The overall system performance can be identified at the top level ability(ies).

The Ability Graph is a qualitative representation of the abilities necessary for vehicle guidance. These abilities can be compared to the required abilities of the current driving situation including a safety margin. The difference is a metric for the risk of operation in the current situation. If the capabilities are sufficient, the situation can be mastered safely. If one or more abilities have a low performance value, the situation can be dangerous and thus, immediate actions to reduce the risk are necessary, either by reducing the "difficulty" of the driving situation or by increasing the vehicle's ability levels.

The self-perception and self-representation and other safety functionalities require additional hard- and software components. These components must not or at most in a tolerable manner affect the functional components of the vehicle guidance system. Functional degradation or self-healing methods can be further safety functionalities.

6.3.3 *Self-Perception*

The self-perception process collects data from all sensors and stores this data for safety purposes like building a self-representation and for functional purposes where the information is used in the functional components. In today's series vehicles self-diagnosis functions are integrated in the components of the vehicle.

Some of the diagnosis functions signalize issues to the driver in the dashboard, others force emergency operation modes of components, e.g., the limp home mode for an engine (Volkswagen 2011). Jerhot et al. (2009) describe an environment perception system with self-diagnosis capabilities. Jerhot et al. (2009) introduce a probabilistic approach for monitoring and an adaptation of the vehicle's functional capabilities. This approach is capable of monitoring track quality, dropouts, tracking time, association success, state prerequisites, distance, azimuth, and speed. It is introduced for the environment perception components of Advanced Driver Assistance Systems. It's applicability to automated vehicles seems possible.

As Dietmayer (2016) points out the testing process of perception systems for automated vehicles has to be done by a mathematical description of so called episodes (a set of situations) and, e.g., Monte-Carlo-Simulations for identifying those critical situations. Field-tests with a large number of driven kilometers are not sufficient because the probability of occurrence is not high enough (Wachenfeld and Winner 2016). Dietmayer also argues that an online monitoring of the perception-performance (consisting of state-, existence-, and classification-uncertainty) is barely possible with state-of-the-art systems. Although a prediction of the performance (in case of sudden failures) is currently impossible. The chapter points out the need for a hardware- and functional-redundant hardware setup for achieving a minimal level of performance in case of sudden failures to fulfill ASIL D requirements. With such setup, a driverless system will not come into (technical) unsolvable situations and can achieve a permanent safe state of operation.

The self-perception is a database of vehicle and system information. Sensors are used to determine vehicle dynamics and actuator values. Additionally, hardware and software heartbeats and cycle times are collected. More sophisticated values are calculated by integrated safety methods in the software components and smaller units. These values are algorithm specific and can only be generated by the functional modules themselves, e.g., a covariance matrix for probabilistic tracking algorithms contains information about the current state of the estimation.

6.3.4 Self-Representation

The self-representation uses the self-perception data and the Ability Graph to determine the current capabilities of the vehicle. Several values from the self-perception are aggregated for providing more complex performance metrics. In this step, the focus of monitoring switches from software and functional modules to the more abstracted ability view. Additionally, a prediction of future capabilities is possible to drive more adapted to the current situation and by that avoid dangerous situations with a high risk level in future maneuvers (Bergmiller 2014; Siedersberger 2003). By comparing current performance and estimating the necessary performance for the current and future situations, system boundaries can be detected and reaching those can be avoided.

The proposed concept should enable automated driving and an early version presented in Reschka et al. (2012a) enabled the first automated driving demonstrations in Germany in the Stadtpilot project in 2010 (Nothdurft et al. 2011).

6.4 Summary and Outlook

This chapter described the process to derive a functional safety concept for automated road vehicles. Additionally we presented the introduction of the Skill Graph in the concept phase and the later transfer and utilization as an Ability Graph in the operation of the vehicle. We expect this concept to work for automated vehicles of SAE Levels 3–5 and will investigate it further in the aFAS project and the Stadtpilot project. Considering functional safety according to the ISO 26262 standard, the components implementing the functional safety concept will gain high ASIL determinations, because of the criticality of automated driving. The Skill Graph enables a safety analysis in the concept phase, which could control complexity by using necessary driving maneuvers as top-level skills and derive the subordinate skills from human tasks.

For the development phase of hardware products, the methods and metrics proposed in the ISO 26262 standard can be applied. For software it is still subject of ongoing research, how correctness of software of control units can be achieved, especially without an extensive testing of the vehicle.

Acknowledgments We would like to thank our project partners in the project Stadtpilot, the project CCC and the project aFAS and our colleagues at the Technische Universität Braunschweig and the project partners in CCC and aFAS for their support of our work. We would like to thank the DFG for funding our research in CCC and the BMWi for funding our research in aFAS.

References

Bagschik, G., Reschka, A., Stolte, T., Maurer, M.: Identification of potential hazardous events for an unmanned protective vehicle. In: Proceedings of the 2016 IEEE Intelligent Vehicles Sym posium (IV), Gothenburg, Sweden, pp. 691–697 (2016)

Bartels, A., Eberle, U., Knapp, A.: AdaptIVe Deliverable D2.1 // System Classification and Glossary (2015)

Benmimoun, M., Pütz, A., Aust, M., Faber, F., Sánchez, D., Metz, B., Saint Pierre, G., Geißler, T., Guidotti, L., Malta, L.: euroFOT SP6 D6.1 Final evaluation results (2012)

Bergmiller, P.: Towards functional safety in drive-by-wire vehicles. PhD Dissertation, Technische Universität Braunschweig (2014)

Dickmanns, E.D.: The 4d–approach to dynamic machine vision. In: Proceedings of the 33rd IEEE Conference on Decision and Control, Lake Buena Vista, pp. 3770–3775 (1994)

Dickmanns, E.D.: The development of machine vision for road vehicles in the last decade. In: Proceedings of the 2002 IEEE Intelligent Vehicle Symposium (IV), Versailles, France, pp. 268–281 (2002)

Dickmanns, E.D.: Dynamic Vision for Perception and Motion Control. Springer, London (2007)

Dickmanns, E.D.: Personal Communication, Braunschweig (2015)

Dietmayer, K.: Predicting of machine perception for automated driving. In: Maurer, M., Gerdes, C., Lenz, B., Winner, H. (eds.) Autonomous Driving – Technical, Legal and Social Aspects, pp. 407–424. Springer, Berlin (2016)

Gasser, T.M., Arzt, C., Ayoubi, M., Bartels, A., Bürkle, L., Eier, J., Flemisch, F., Hesse, T., Huber, W., Lotz, C., Maurer, M., Ruth-Schumacher, S., Schwarz, J., Vogt, W.: Rechtsfolgen zunehmender Fahrzeugautomatisierung: gemeinsamer Schlussbericht der Projektgruppe. Wirt schaftsverlag NW, Verlag für neue Wissenschaft (2012)

Gerdes, C., Thornton, S.M.: Implementable ethics for autonomous vehicles. In: Maurer, M., Gerdes, C., Lenz, B., Winner, H. (eds.) Autonomous Driving – Technical, Legal and Social Aspects, pp. 87–102. Springer, Berlin (2016)

Grunwald, A.: Societal risk constellations for autonomous driving. Analysis, historical context and assessment. In: Maurer, M., Gerdes, C., Lenz, B., Winner, H. (eds.) Autonomous Driving – Technical, Legal and Social Aspects, pp. 641–663. Springer, Berlin (2016)

Hörwick, M.: Sicherheitskonzept für hochautomatisierte Fahrerassistenzsysteme. PhD Dissertation, Technische Universität München (2011)

Hörwick, M., Siedersberger, K.H.: Strategy and architecture of a safety concept for fully automatic and autonomous driving assistance systems. In: Proceedings of the 2010 IEEE Intelligent Vehicles Symposium (IV), San Diego, pp. 955–960 (2010)

Huang, H. (ed.): Autonomy Levels for Unmanned Systems (ALFUS) Framework – Volume I: Terminology – Version 2.0. NIST Special Publication 1011-I-2.0 (2008)

Huang, H., Messina, E., Albus, J.: Autonomy Levels for Unmanned Systems (ALFUS) Framework – Volume II: Framework Models – Version 1.0. NIST Special Publication 1011-II-1.0 (2007)
ISO 26262:2011: Road Vehicles – Functional Safety. ISO, Geneva (2011)
Jerhot, J., Form, T., Stanek, G., Meinecke, M., Nguyen, T., Knaup, J.: Integrated probabilistic approach to environmental perception with selfdiagnosis capability for advanced Driver Assistance Systems. In: 12th International Conference on Information Fusion (FUSION '09), Seattle, pp. 1347–1354 (2009)
Kriso, S., Hamann, R., Gebauer, C.: Die Item Definition der ISO 26262 – Unangenehme Auswirkungen bei ungeschickter Wahl der Systemgrenze. In: VDI-Berichte Nr. 2188 (2013)
Lin, P.: Why ethics matters for autonomous cars. In: Maurer, M., Gerdes, C., Lenz, B., Winner, H. (eds.) Autonomous Driving – Technical, Legal and Social Aspects, pp. 69–85. Springer, Berlin (2016)
Mann, M.: The car that drives itself. Popular Sci. 172, 75–79, 226–227
Matthaei, R.: Personal Communication, Braunschweig (2014)
Matthaei, R.: Wahrnehmungsgestützte Lokalisierung in fahrstreifengenauen Karten für Fahrerassistenzsysteme und automatisches Fahren in urbaner Umgebung. PhD Dissertation, Technische Universität Braunschweig (2015)
Matthaei, R., Maurer, M.: Autonomous driving – a top-down approach. Automatisierungstechnik – Auto. **63**(4), 155–167 (2015)
Matthaei, R., Reschka, A., Rieken, J., Dierkes, F., Ulbrich, S., Winkle, T., Maurer, M.: Autonomous driving. In: Winner, H., Hakuli, S., Lotz, F., Singer, C. (eds.) Handbook of Driver Assistance Systems, pp. 1519–1556. Springer, Cham (2016)
Maurer, M.: Flexible Automatisierung von Straßenfahrzeugen mit Rechnersehen. VDI-Verlag (2000)
Maurer, M.: Forward collision warning and avoidance. In: Eskandarian, A. (ed.) Handbook of Intelligent Vehicles, pp. 657–687. Springer, London (2012)
Nagel, H., Enkelmann, W.: Generic road traffic situations and driver support systems. In: Proceedings of the 5th Prometheus Workshop. Munich, Germany, pp. 76–85 (1991)
National Highway Traffic Safety Association (NHTSA): Preliminary statement of policy concerning automated vehicles (2013)
Nothdurft, T., Hecker, P., Ohl, S., Saust, F., Maurer, M., Reschka, A., Böhmer, J.R.: Stadtpilot: first fully autonomous test drives in urban traffic. In: 2011 IEEE International Annual Conference on Intelligent Transportation Systems (ITSC). Washington, DC, pp. 919–924 (2011)
Ohl, S.: Fusion von Umfeld wahrnehmenden Sensoren in städtischer Umgebung. PhD Dissertation, Technische Universität Braunschweig (2014)
Ohl, S., Maurer, M., Häusler, K., Holldorb, C.: Autonomes Fahren im Strassenbetriebsdienst. In: 13. Braunschweiger Symposium Automatisierungssysteme, Assistenzsysteme und einge bettete Systeme für Transportmittel (AAET), Brunswick, Germany (2012)
Pellkofer, M.: Verhaltensentscheidung für autonome Fahrzeuge mit Blickrichtungssteuerung. PhD Dissertation, Universität der Bundeswehr München (2003)
Rasmussen, J.: Skills, rules, and knowledge; signals, signs, and symbols, and other distinctions in human performance models. IEEE Trans. Syst. Men Cybern. **13**(3), 257–266 (1983)
Rauskolb, F.W., Berger, K., Lipski, C., Magnor, M., Cornelsen, K., Effertz, J., Form, T., Graefe, F., Ohl, S., Schumacher, W., Wille, J.M., Hecker, P., Nothdurft, T., Doering, M., Homeier, K., Morgenroth, J., Wolf, L., Basarke, C., Berger, C., Gülke, T., Klose, F., Rumpe, B.: Caroline: an autonomously driving vehicle for urban environments. J. Field Rob. **25**(9), 674–724 (2008)
Reschka, A.: Safety concept for autonomous vehicles. In: Maurer, M., Gerdes, C., Lenz, B., Winner, H. (eds.) Autonomous Driving – Technical, Legal and Social Aspects, pp. 473–496. Springer, Berlin (2016)
Reschka, A., Maurer, M.: Conditions for a safe state of automated road vehicles. Inf. Technol. **57**(4), 215–222 (2015)
Reschka, A., Böhmer, J.R., Gacnik, J., Köster, F., Wille, J.M., Maurer, M.: Development of software for open autonomous automotive systems in the Stadtpilot project. In: Proceedings of the 8th International Workshop on Intelligent Transportation (WIT 2011), Hamburg, Germany, pp. 81–86 (2011)

Reschka, A., Böhmer, J.R., Nothdurft, T., Hecker, P., Lichte, B., Maurer, M.: A surveillance and safety system based on performance criteria and functional degradation for an autonomous vehi cle. In: Proceedings of the 2012 IEEE International Annual Conference on Intelligent Trans portation Systems (ITSC), Anchorage, pp. 237–242 (2012a)
Reschka, A., Böhmer, J.R., Saust, F., Lichte, B., Maurer, M.: Safe, dynamic and comfort-able longitudinal control for an autonomous vehicle. In: Proceedings of the 2012 IEEE Intelligent Vehicles Symposium (IV), Alcalá des Henares, Spain, pp. 346–351 (2012b)
Reschka, A., Bagschik, G., Ulbrich, S., Nolte, M., Maurer, M.: Ability and skill graphs for system modeling, online monitoring, and decision support for vehicle guidance systems. In: Proceedings of the 2015 IEEE Intelligent Vehicles Symposium (IV). Seoul, Korea, pp. 933–939 (2015)
Schopper, M., Henle, L., Wohland, T.: DISTRONIC PLUS mit lenk-assistent und Stop&Go pilot. ATZextra. **18**(5), 106–114 (2013)
Schuldt, F., Lichte, B., Maurer, M., Scholz, S.: Systematische Auswertung von Testfällen für Fahrfunktionen im modularen virtuellen Testbaukasten. In: Workshop Fahrerassistenzsysteme, Walting, Germany, pp. 169–179 (2014)
Siedersberger, K.H.: Komponenten zur automatischen Fahrzeugführung in sehenden(semi-) autonomen Fahrzeugen. PhD Dissertation, Universität der Bundeswehr München (2003)
Siedersberger, K.H., Gregor, R., Pellkofer, M.: Diskussionsrunden der Arbeitsgruppe Verhalten, Universität der Bundeswehr München (2000)
Smith, B.W.: Lawyers and engineers should speak the same robot language. In: Calo, R., Froomkin, A.M., Kerr, I. (eds.) Robot Law, pp. 78–101. Edward Elgar, Cheltenham (2016)
Society of Automotive Engineers (SAE): Taxonomy and Definitions for Terms Related to On-Road Motor Vehicle Automated Driving Systems (J3016) (2014)
Stolte, T., Reschka, A., Bagschik, G., Maurer, M.: Towards automated driving: unmanned safe guarding vehicle for highway hard shoulder roadworks. In: Proceedings of the 2015 IEEE International Annual Conference on Intelligent Transportation Systems (ITSC), Las Palmas, Spain, pp. 672–677 (2015a)
Stolte, T., Bagschik, G., Reschka, A., Maurer, M.: Automatisch fahrerlos fahrendes Absicherungs fahrzeug für Arbeitsstellen auf Autobahnen (aFAS). In: AAET – Automatisierungssysteme, Assistenzsysteme und eingebettete Systeme für Transportmittel, Braunschweig, Germany (2015b)
Tölle, W.: Ein Fahrmanöverkonzept für einen menschlichen Kopiloten. VDI-Verlag (1996)
Tsugawa, S.: Vision-based vehicles in Japan: machine vision systems and driving control systems. IEEE Trans. Indus. Electron. **41**(4), 398–405 (1994)
VDI 2206: VDI-Richtlinien – Entwicklungsmethodik für mechatronische Systeme (2004)
Volkswagen, A.G.: Volkswagen Passat B7 Betriebsanleitung (2011)
Wachenfeld, W., Winner, H.: The release of autonomous vehicles. In: Maurer, M., Gerdes, C., Lenz, B., Winner, H. (eds.) Autonomous Driving – Technical, Legal and Social Aspects, pp. 425–449. Springer, Berlin (2016)
Wachenfeld, W., Winner, H., Gerdes, C., Lenz, B., Maurer, M., Beiker, S.A., Fraedrich, E., Winkle, T.: Use cases for autonomous driving. In: Maurer, M., Gerdes, C., Lenz, B., Winner, H. (eds.) Autonomous Driving – Technical, Legal and Social Aspects, pp. 9–37. Springer, Berlin (2016)
Wille, J.M.: Manöverübergreifende autonome Fahrzeugführung in innerstädtischen Szenarien am Beispiel des Stadtpilotprojekts. PhD Dissertation, Technische Universität Braunschweig (2012)
Wille, J.M., Saust, F., Maurer, M.: Stadtpilot: driving autonomously on Braunschweig's inner ring road. In: 2010 IEEE Intelligent Vehicles Symposium (IV), San Diego, pp. 506–511 (2010)
Winkle, T.: Development and approval of automated vehicles: considerations of technical, legal, and economic risks. In: Maurer, M., Gerdes, C., Lenz, B., Winner, H. (eds.) Autonomous Driving – Technical, Legal and Social Aspects, pp. 589–618. Springer, Berlin (2016)

Chapter 7
A Method for an Efficient, Systematic Test Case Generation for Advanced Driver Assistance Systems in Virtual Environments

Fabian Schuldt, Andreas Reschka, and Markus Maurer

Abstract In this chapter, a method for an efficient, systematic test case generation for the test of advanced driver assistance systems in virtual environments is presented. The method is one of four steps in a systematic test process. These four steps are (1) analysis of the system, (2) test case generation, (3) test execution, and (4) test evaluation. The analysis serves to identify factors that have an impact to the system. The aim of the test case generation is to discretize value-continuous parameters into equivalence classes and to reduce the number of test cases for necessary test coverage. The test case generation uses combinatorial algorithms to achieve this objective. A test case is generated based on a 4-level model, which consists of the road network, adjustments for special situations, dynamic elements, and environmental conditions. To generate reproducible test cases, a special control for dynamic elements is introduced to adapt the behavior of dynamic elements to non-deterministic target elements. The test case generation is presented in a case study of a constriction assist. The test evaluation is used to verify the system and to replay test cases or important factors to the previous steps of the test concept.

Keywords Systematic test case generation • 4 level scenario model • Virtual environments • Combinatorial test case generation • Automated driving functions

7.1 Introduction

7.1.1 Motivation

Testing of software-intensive systems is an important step in the development process of distributed software systems. The V-model (Rook 1986; Sommerville

---

F. Schuldt (✉)
Peiner Straße 21, 38536 Meinersen, Germany
e-mail: fabian.schuldt1@gmail.com

A. Reschka • M. Maurer
Institut für Regelungstechnik, Hans-Sommer-Str. 66, Braunschweig 38106, Germany

H. Winner et al. (eds.), *Automotive Systems Engineering II*,
DOI 10.1007/978-3-319-61607-0_7

2007) and the ISO 26262 standard (ISO 26262 2011, Part 6) define a process for the development of requirements for a system, test cases, the software architecture (integration), and software units on the left side of the V-model. On the right side of the V-model, test cases are executed to test every requirement on the level of software units, software integration, and system tests. Only if all requirements are tested with test cases and the system is verified for a required quality, the system can be released. To ensure that the system is properly tested, a sufficient number of test cases on every level on the right side of the V-model have to be executed. Determining the necessary number of test cases for a sufficient test of a system is still an open research question.

In safety-relevant systems, faults, which result in harm to the environment of the system or the system itself, are not accepted at any time during operation. Thus, it is challenging to test the system sufficiently and to generate a high test coverage, which is needed for a release of the system and an assumption that the system operates as reliable as technologically possible in the target environment. A high test coverage or test depth can only be reached with a large number of test cases, because theoretically all combinations of impact factors have to be tested. If the system has value-continuous impact factors, an infinite number of test cases can be generated and complete test coverage of the system is not even theoretically reachable. In this case, the values of impact factors have to be discretized to reduce the large number of test cases to an economically feasible number.

By a discretization of parameter values, not every possible value for each parameter is tested. Thus, a fault probability for the system remains. For a release of the system, it has to be evaluated, if the remaining fault probability is acceptable for the market and the society.

Only a limited number of test cases can be run, because every test case is associated with costs and time. If not all necessary test cases, which are required to test all requirements of the system, can be executed, the release of the system can be harmful, because not all requirements are tested. Thus, the hazards and the remaining fault probability for a release of the system have to be estimated. Otherwise the system must not be released. Therefore, it is important that all executed test cases are efficient and have a high coverage of the complete test requirements, which are defined for the system.

Thus, an increasing amount of time has to be spent for a large number of test cases in the development process. Therefore, the costs for testing are often higher than 50% of the complete development cost budget. The portion of testing costs can even be higher for safety-relevant systems (Ammann and Jeff 2008).

Examples for software-intensive systems are advanced driver assistance systems, which have to assist the driver when conducting the vehicle. Currently available systems are adaptive cruise control (ISO 22179 2009; ISO 15622 2010) and lane keeping assistance systems (ISO 11270 2014) for instance. The systems are using more and more lateral and longitudinal interventions to the vehicle to assist the driver. Other systems take over a part of the driving task. Therefore, incorrect interventions to the lateral and longitudinal control of the vehicle can have devastating consequences. For that reason, it has to be secured that only a sufficient-tested software release is used in vehicles on public roads. The software release has to have a reliable behavior at every time, which means that faulty interventions are also prevented out of the system boundaries for instance. Again, this is only possible with a high test depth, which can only be generated with a sufficient number of test cases; each of them tested with one or more suitable test methods. Finding a suitable test method for different systems is also an open research question.

Advanced driver assistance systems currently available are predominantly designed for non-urban environments. A lane keeping assistance system is only active while driving with a speed over 72 km/h (ISO 11270 2014). Consequently, the system may only be used on highways or rural roads. An application is currently not possible in urban environments. Current research projects like UR:BAN[1] (Scholl 2015) are developing advanced driver assistance systems for urban environments to reduce the number and the severity of accidents. The urban environment imposes higher requirements on driver assistance systems by requiring collision avoidance in traffic with lower time-to-collision to a large number of moving objects around the ego vehicle or a lower time-to-lane-crossing. Additionally, the traffic flow is considerably denser and the surrounding environment is not well-structured as on highways or rural roads. Besides merging in and out vehicles, driver assistance systems have to detect turning vehicles and oncoming traffic. Due to possibly conflicting intentions, intersection scenarios may result in a particular high level of complexity. Thus, urban environments describe complex scenarios for advanced driver assistance systems, because a large amount of factors, which can be varied in a high value range, have an impact to the system.

To manage the complexity of scenarios in urban environments for a test of advanced driver assistance systems, a systematic and efficient test case generation is required.

7.1.2 Related Work

Today, advanced driver assistance systems are already tested in different ways. This chapter presents the test case derivation and the generation in other approaches with black-box testing methods.

In Schmidt (2012), an alternative approach for the generation of test scenarios for advanced driver systems is described. The main focus is the generation of scenarios for camera-based systems and therefore, the generation of photo-realistic pictures. Schmidt uses combinatorial methods for the generation of test cases, which are based on scenery tiles. The combinatorial methods are used for a scenario generation on the meta-levels of *track*, *environment*, *surrounding*, and *test-relevant object*. However, the methods are not used inside of the meta-levels, like in the

[1]In German: Urbaner Raum: Benutzergerechte Assistenzsystem und Netzmanagement (Urban Space: User oriented assistance systems and network management).

approach in this chapter. Thus, faults can only be detected on the level of the meta-levels and not on the basis of single parameter values.

Schmidt also suggests the generation of equivalence classes and a boundary value analysis for generating test cases. The discretization of value-continuous parameters into equivalence classes and the risks and hazards of testing with equivalence classes are not discussed in Schmidt (2012).

In Eltaher (2013), a cognitive-oriented test approach is presented. The main idea of the approach is to study the test method of test experts and to teach an expert test system, which can test the system efficiently and effectively and generate new test cases automatically.

Expert tests are an additional test method to the systematic testing, which is presented in this chapter. In some cases, experts can also find faults in the system after a systematic testing, because they know the critical test cases through their expertise about the systems. However, experts often cannot explain their test methods, because the knowledge about the test method and the test case generation exits only in an implicit manner. Eltaher uses situation-operator-models to model the implicit knowledge of experts for an expert test system. In a case study with an infotainment system, Eltaher demonstrates that a trained expert system can find more faults in the system than a human tester. The tested infotainment system is a system with discrete input parameters. Thus, a discretization of the parameter values is not necessary for the test case generation and therefore, the problem of performing a discretization has not explicitly been discussed.

However, the approach is also interesting for integration into a systematic testing to discretize value-continuous parameters. Experts discretize value-continuous parameters implicitly in real world tests of advanced driver assistance systems by the choice of the test scenarios and test environment. Currently, only the explicit knowledge of test experts may be used for the discretization of value-continuous values in systematic testing. With the approach of Eltaher, the possibility of the integration of the implicit knowledge may be given. The use of the approach in the systematic testing has to be proved in a case study or future research work.

In Lindlar (2012), an approach of a model-based evolutionary test method is presented. The test method is demonstrated by means of an adaptive cruise control system. The input parameters are the *velocity of the vehicle in front of the ego-vehicle*, *the driver inputs to the system*, and *the curvature of the lane*. The results of the test case execution are evaluated by a fitness function, whereby a high fitness shows a good test result. In a number of iterations, a global minimum for the fitness function is searched with an evolutionary algorithm. The global minimum is likewise the most critical test case.

An advantage of the approach is the not needed prior discretization of value-continuous parameters for the test case generation. For value-continuous parameters, the evolutionary algorithm searches the most critical value between a minimal and maximal value during the test process. Thus, the approach can identify faults in the system without any prior information. The challenge of this approach is to find a suitable fitness function for the system. In the case study with one vehicle in front of the ego-vehicle, the fitness depends only on five parameters. In scenarios with more vehicles around the ego-vehicle and different environments, determining a suitable fitness function is more challenging than finding faults in the system.

Additionally, the evolutionary algorithm optimizes the fitness function in the direction of the global optimum. Thereby, only single faults can be detected with the algorithm. Also, interactions between different parameters cannot be detected. A screening test about the complete parameter range is not possible, which is however with combinatorial testing.

In Hilf et al. (2010), an approach of a test case generator for discrete state scenarios is presented. The approach is demonstrated in the case study for a crosswind stabilization function with the input parameters *bank angle of the road*, *wind properties*, and vehicle inputs like *acceleration*, *brake pressure*, and *steering angle*. Based on the input parameters and all previously generated test cases, the test case generator generates new test cases to test all possible system states by at least one single test case. The test strategy is compared to a chess game. In the game, new scenarios are identified to achieve non-tested system states. The results of the test case execution are compared to predefined conditions for the output parameters. The test coverage is measured by the number of reached system states. If all system states are well-known, the approach is a suitable test method to find faults in the system.

As system tests of advanced driver assistance systems are mostly black-box tests, most parts of the internal structure and the system states are unknown. Hence, it is not possible to modify the scenarios to find a test case for all system states. Additionally, discrete parameters are required for the test case generation. Again, the discretization of value-continuous parameters into value-discrete parameters is a challenge for this method.

7.2 The Efficient, Systematic Test Method in Virtual Environments

This section presents the components of the efficient, systematic test method. The main goal of this method is to generate and execute tests to create a sufficient test depth for driver assistance systems in different environments like highways, rural roads, or urban environments. The method is efficient and systematic by the generation of test cases for the system.

Efficient in this context means that costs and time for the tests can be reduced with the method compared to multi-million kilometer field tests. This will be reached with an analysis of the system for impact factors like in standards and guidelines for highways, rural roads, and urban roads (FGSV 1980, 1995, 1996, 2006, 2008), to generate pointedly representative test cases for the respective driving function for different environments. This means for the test that the scenarios with standard parameters, which can be discovered by the standards and guidelines, are tested. Here, this means that the driving function is tested for an operation on all highways, rural roads, and urban roads, which are constructed according to these standards and guidelines.

Systematic in this context means that the test cases are generated on the basis of a unified 4-level model. With this model, a flexible generation of test cases is possible. Additionally, combinatorial algorithms are used to reduce the number of test cases by well-defined test coverage (see sub-chapter combinatorial test case generation). The requirements for the test case generation are described in detail in the next sub-chapters.

The deficit of this test method is that scenarios in the real world exist, which are not tested when these are outside of the standards. To identify these scenarios, alternative analysis methods have to be found. Additionally, there is a deficit, if no standards and guidelines are available for the country or the environment, where the system under test has to operate. In this case, the method would also fail.

The test method is designed as a system test in the V-model. The system under test is regarded as a black-box for the method. Thus, the system is only tested on the basis of the input and output parameters.

The method generates test cases for a test in a simulation environment. Testing with the simulation has some advantages like testing faster than in real time, running multiple test cases in parallel, the reproducibility of tests, when all elements of the simulation are deterministic, and also the high flexibility of tests. Thereby, with the simulation the systems can be tested efficiently, because costs and time can be controlled by the number of test runs.

To test advanced driver assistance systems, the perception algorithms and the driving function of the system have to be tested.

For testing perception algorithms, sensor models have to be implemented in the simulation to generate sensor raw data. But these models can only simulate a part of the effects, which happen to sensors in the real world. For instance, to simulate the interactions between environmental and lighting conditions, complex models with many parameters are necessary for an exact simulation of the effects on camera systems. Generating such complex sensor models is maybe not possible in the simulation, because there is no method to identify all impact parameters for a model, which can generate realistic sensor raw data in all possible situations. The sensor models can only simulate a defined part of the sensor effects.

Thus, the possibility to verify perception algorithms in a simulation environment can be limited, due to missing models for the environment, resulting in a difficult generation of realistic sensor raw data. Real world tests are more suitable for these tests, because all sensor effects occur in the real world.

The driving function receives data about the environment from the perception by interfaces, e.g. a scene (Ulbrich et al. 2015). This interface data (e.g. the position and velocity of the surrounding vehicles) can be perfectly generated with the simulation environment in different qualities from perfect to biased data. Due to the reproducibility and flexibility of the simulation, the driving function can be tested in a short time with perfect conditions in a large amount of different scenarios without excessive efforts. This is not possible in real world tests, because all tests can only be executed in real time and parallelization is only possible with multiple

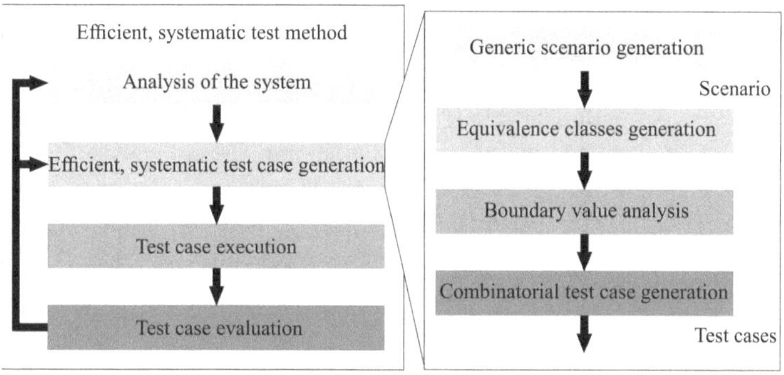

Fig. 7.1 The four steps of the efficient, systematic test method with a detailed view into the efficient, systematic test case generation

vehicles and test drivers. Thus, the simulation is a useful tool for the test of a driving function.

Due to the advantages of the simulation and the described possibilities of testing in the simulation, the efficient, systematic test method aims to test driving functions primarily in the simulation. The perception algorithms, which process the sensor raw data, are currently not tested.

Figure 7.1 illustrates the steps of the efficient test method with a detailed view into the test case generation.

The method can be divided into four steps: (1) analysis of the system, (2) efficient, systematic test case generation, (3) test case execution, and (4) the test case evaluation (see Fig. 7.1), which are described in the following:

1. *Analysis of the system*:
   In the first step, the advanced driver assistance system has to be analyzed to identify parameters, which have an impact in the application scenarios. Therefore, system requirements, scenario catalogues, or standards and guidelines are used to identify those major impact factors. Additionally, test experts can be interviewed to identify significant impact factors. Another approach to identify impact factors for the driving function based on an ability graph is described in Chap. 6 in this book.
2. *Efficient, systematic test case generation*:
   In the second step, the test cases are generated with the efficient systematic test case generation, based on the analysis of the system for impact factors. Blackbox testing methods are used to discretize the value-continuous impact parameters into discrete values and equivalence classes. Combinatorial testing methods are used to reduce the number of test cases in a test suite, which is a collection of all test cases. A single test case is described on the basis of the 4-level model. The single steps of the efficient systematic test case generation are described in detail in the following chapter.

3. *Test case execution*:
   The test case execution is performed in the simulation to use the advantages of the simulation. The degree of the simulation is variable on the different levels of the test case execution to stimulate the system with information in different qualities. The test cases are executed on the simulation levels *software-in-the-loop*, *driver-in-the-loop*, *vehicle-hardware-in-the-loop* (Verhoeff et al. 2000; Gietelink et al. 2003; Hendrik 2010), and *vehicle-in-the-loop* (Bock et al. 2007; Bock 2008). Additionally, the test case execution provides a method for a mapping of these test cases to the different simulations (Schuldt et al. 2015).
4. *Test case evaluation*:
   The test case evaluation is the last step in the efficient systematic test method. To compare the results of the test cases, evaluation methods of the quality assurance, like the quality loss function, are used (Taguchi et al. 2007). With these, results of the test case execution can be compared on the basis of different criteria and rated with different metrics. The next step in the test case evaluation is the interpretation of the test results. It should be analyzed, which parameters and which numeric parameter values have a significant impact on the evaluation criteria. Different statistic tools can be used, like the analysis of mean values (Siebertz et al. 2010) or the analysis of variance (Taguchi et al. 2007). With these tools, the impact of each parameter can be estimated. The results can also be used for a further iteration of the test steps with the aim to identify new impact factors or new discrete values for continuous variables.

7.3 Requirements on an Efficient, Systematic Test Case Generation

This chapter describes the requirements on an efficient test case generation. For a complete test of the advanced driver assistance system, all combinations of impact factors have to be tested. The execution of this exhaustive number of tests is not possible in reality.

Therefore, a test case generation method is needed, which generates test cases efficiently and systematically. This test case generation method has to generate test cases, which fulfill the following requirements:

- *Non-redundant*:
  No redundant test case shall be generated. This means that no test cases with the same parameter value combination of dependent parameters should be executed with more than one test case. Redundant test cases will only cost and waste valuable test time without increasing the test coverage. So, it is important for an efficient test case generation to minimize the number of redundant test cases.
- *Representative*:
  The generated test cases have to be representative for the system under test. There must be the correct parameters and their values chosen in the test cases to generate representative test cases on the system requirements.

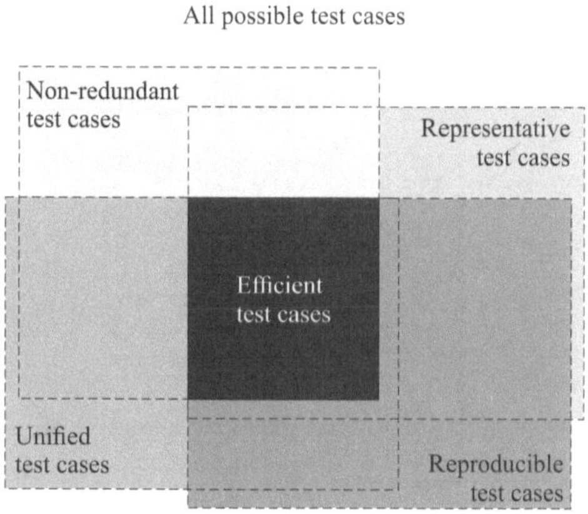

Fig. 7.2 Efficient test cases are the intersection of test cases, which are non-redundant, representative, unified, and reproducible

- *Unified*:
  The generated test cases need a unified description for different scenarios, including different environments, like highways, rural roads, or urban roads, and additionally pass and fail criteria for the test cases. An adaption of the test cases to new test objects has to be possible without a large effort. All identified impact parameters have to be presented in an unified structure. Additionally, the structure should be extendable, if new parameters are identified. Parameter values have to be varied easily to generate different test cases.
- *Reproducible*:
  The generated test cases have to be reproducible, especially if non-deterministic elements, like a real driver or a real vehicle, are part of the test case. The elements in the test case have to adapt their behavior to the behavior of the test object to generate reproducible test cases. Thus, specific models for the behavior of elements of the test case are necessary.

Only when these four requirements are fulfilled by the generated test cases, the test case generation is efficient and systematic. Figure 7.2 illustrates the presented requirements on the test case generation.

7.4 Unified Scenario Generation for Efficient Test Cases on the Basis of the 4-Level Model

This section describes the single steps of the unified scenario generation for advanced driver assistance systems. The scenario generation bases on a 4-level model. A scenario can be flexibly assembled on a selection of one or more of the four levels. E.g. static scenarios without dynamic elements can be defined on the basis of the first two levels. On every level of the scenario generation, the principle of the systematic test case generation is applied, which is described in detail in the

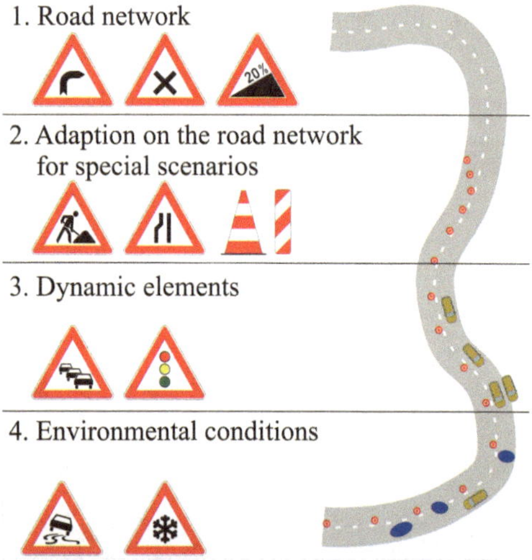

Fig. 7.3 The 4-level model for a unified scenario generation

second part of this sub-chapter. Figure 7.3 shows the structure of the unified scenario generation on the basis of the 4-level model, which is described in the following.

7.4.1 Level 1: Road Network

On the first level of the unified scenario generation, the road network has to be defined. Therefore, the geometry and topology of the roads have to be specified. For this, the basic elements for road design *straight lines, curves*, and *clothoids* are used. Straight lines are described only by a length. Curves are defined by a constant radius or curvature, which is unequal to zero, and a length. To combine straight lines with curves, clothoids are used. Clothoids are defined by a start curvature, an end curvature, and a length. Therefore, clothoids represent transition elements between straight lanes and curves or curves with different curvatures.

The geometries of the road network are extracted from current standards and guidelines for the creation of highways, rural roads, and urban roads (FGSV 1980, 1995, 1996, 2006, 2008). These geometric parameters can be, for example, the allowed minimal and maximal curvature and/or lateral and longitudinal elevation profiles. Table 7.1 shows some exemplary values from the construction guidelines of a highway in Germany.

If the road cannot be described with these three basic elements, it is also possible to define different splines to describe the road. Therefore, it is also possible to create and define non-standard roads.

After defining the geometries of the road, the topology elements of the road are described. Therefore, the number of lanes has to be defined, along with their width

Table 7.1 Overview of design parameters of different highways in Germany (FGSV 2008)

	EKA 1A	EKA 1b	EKA 2	EKA 3
Road function	Highway	Over regional highway	Highway similar road	Urban highway
Maximal length of a straight line [m]	2000	–	–	–
Minimal curve radius [m]	900	720	470	280
Minimal clothoid parameter $A = \sqrt{R \cdot L}$ [m]	300	240	160	90

EKA in German: "Entwurfsklasse" (draft class)
R Radius, *L* Length of the element

Table 7.2 Overview of typical cross sections in Germany (FGSV 2008)

Typical cross section	Border [m]	Hard shoulder [m]	Margin strip [m]	Driving lane [m]				Hard shoulder [m]	Central reserve [m]
				First	Second	Third	Fourth		
RQ25	1.5	2	0.5	3.5	3.25			0.5	2.5
RQ28	1.5	2.5	0.75	3.5	3.5			0.75	3.0
RQ31	1.5	3	0.75	3.75	3.75			0.75	4.0
RQ36	1.5	2.5	0.5	3.75	3.5	3.5		0.75	4.0
RQ43.5	1.5	2.5	0.5	3.75	3.75	3.5	3.5	0.75	4.0

RQ: German: "Regelquerschnitt" (cross section)

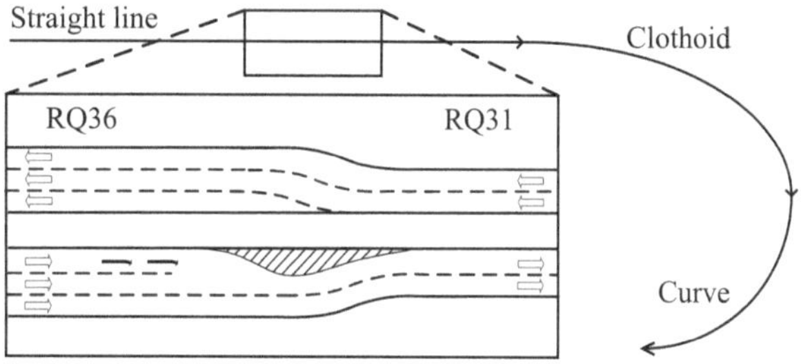

Fig. 7.4 An exemplary basic road, consisting of a straight line, clothoid, curve, and a transition between the typical cross section RQ36 and RQ31 on the straight line

and markings. The lane markings have to be specified by a width, color, and also a style, like solid or broken. Additionally, the state of the road surface has to be specified, like wetness, dirt, or the abrasion of the road. In the standards and guidelines, different typical cross sections are available for the road topology. These describe the range of allowed lane widths in relation to the road type, for instance. Table 7.2 gives an overview of some typical cross sections in Germany.

Figure 7.4 shows the route of an example road, consisting of lines, curves, and clothoids with two typical cross sections for a highway. The figure also shows a transition between the two different typical cross sections, which is also defined in the standards and guidelines.

The basic roads can also be combined to intersections. Therefore, the standards and guidelines for the creation of urban roads are used, where default intersection topologies are defined (FGSV 2006). To describe an intersection, the number of incoming and outgoing roads, and the connecting lanes between these roads have to be defined. Furthermore, the number of lanes for left and right turning has to be specified, along with their lane markings.

7.4.2 Level 2: Adaption of the Road for Special Situations

On the second level, situation-dependent adaptions, which are required for special applications or advanced driver assistance systems with special requirements, are added to the basic road.

These adaptions are for example the surroundings of the road or static elements on the road. If it is important for the test that the basic roads or intersections are in an urban environment, surrounding elements like houses, street lights, and traffic lights have to be added to the basic roads to generate an urban environment.

Examples for two generated intersections in an urban environment are illustrated in Fig. 7.5.

Possible adaptions are additional roads, which are required (e.g. to define road works). These adaptions are required for a test of the constriction assist, which is developed in the research project UR:BAN (Scholl 2015). The road works are also generated according to the current standards and guidelines, like the RSA in Germany (FGSV 2009). Figure 7.6 illustrates a roadwork zone according to the guidelines.

The left lane of the road is closed through a lateral barrier. The barrier is specified by a length and a lateral offset. The ratio between the length and the offset is defined for highways, rural roads, and urban roads. According to the offset of the lateral barrier, the right lane has a remaining width of the delta of the road width with two lanes and the lateral offset. The standards and guidelines provide a minimal width of the right lane in such roadwork zones. For protecting the roadwork zone, additional lane markings with different colors and static elements, like traffic cones, traffic beacons, or concrete barriers are added to the roadwork zone. The distance of the roadwork elements in lateral and longitudinal direction can be varied. The basic road in the first level of the 4-level model is under the adapted road, which means that its geometry is still existent and visible, but may not be used in the construction zone. Thus, the lane markings of the basic road can be used as inconsistent information for systems under test.

Fig. 7.5 Examples of two intersections defined on the first and second level of the model

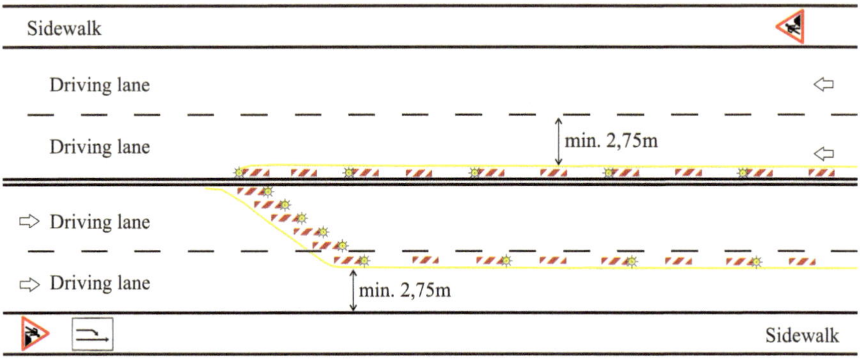

Fig. 7.6 A roadwork zone according to the current guideline on an urban road

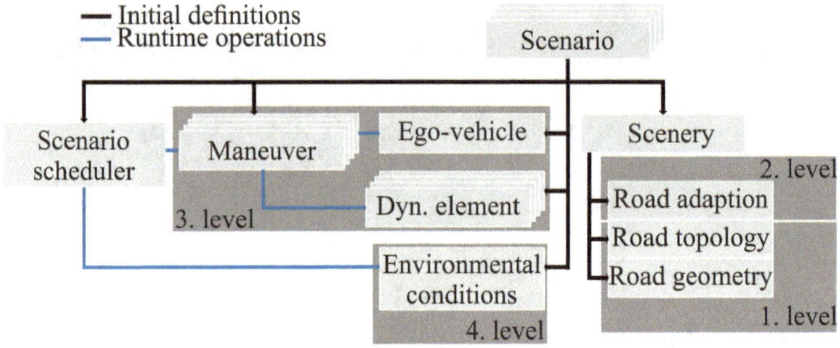

Fig. 7.7 Structure of the control for the dynamic elements and the first and second level of the unified scenario generation in the context of the scenario

7.4.3 Level 3: Dynamic Elements

After defining the basic track and optional adaptions, the quantity and the behaviour of the dynamic elements are defined on the third level. Figure 7.7 illustrates the integration of the first two levels of the 4-level model in the context of a test scenario.

The dynamic elements can be defined on various abstractions levels. On the top level, it should be possible to define dynamic elements with information about the traffic flow. Thereby, it is possible to generate random traffic situations on a defined road. For the test of advanced driver assistance systems, this abstraction layer is useful to generate random scenarios without any detailed information about test cases. The traffic can be described by little information, like the number of other road-users. However, on this level it is not possible to create reproducible traffic scenarios. To generate reproducible scenarios, a lower abstraction level of the definition of dynamic elements is also provided in the unified scenario generation.

On the lower abstraction level, it is possible to define the traffic on a higher specific level, like following a vehicle, parallel driving vehicles, or lane changing vehicles, based on different parameters.

To generate reproducible scenarios, it is important that the scene as perceived by the ego-vehicle is still the same. This means that the other vehicles in the scenario have to react to the actions of the ego-vehicle. For example, if the ego-vehicle is driving with different velocities in the scenario, the other vehicles have to adapt their own behavior to the velocity of the ego-vehicle. Only if this requirement is fulfilled, reproducible scenarios can be generated and efficient test cases with dynamic elements can be executed.

To implement this in the simulation, the scenario generation uses a special control of the dynamic elements. Figure 7.7 illustrates the structure of the control for the dynamic elements in the context of the scenario.

On the right side of the figure, the stationary elements of the scenario are shown, which are also called scenery (Geyer et al. 2014). These elements are described on the first two levels of the unified scenario generation. The left side of the figure shows the dynamic elements, which are the ego-vehicle or the other driving vehicles for example. To control the dynamic elements, different maneuvers are defined in the scenario, which are applied to the dynamic elements by a scenario scheduler. The scenario scheduler has also the option to control the environmental conditions, which are described in the next sub-chapter. With the information about the ego-vehicle, the scheduler has the opportunity to adjust the maneuver to the behavior of the ego-vehicle to generate reproducible scenarios from the viewpoint of the ego-vehicle.

Every maneuver has also an internal structure, which is illustrated in Fig. 7.8.

A maneuver has one or more start conditions. When one of these conditions is fulfilled, the scheduler of the scenario starts the maneuver. For instance, the start conditions are a relative distance or relative velocity to the ego-vehicle, or a specific location on the road, which can be a curve with a defined curvature. Thereby, it is possible to generate special situations at different locations, which can be critical situations for the system under test.

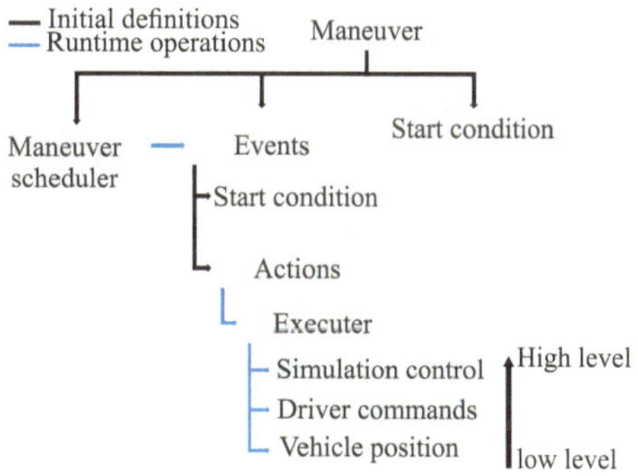

Fig. 7.8 Structure of a maneuver with events and actions

Every maneuver has its own scheduler, which has the task to control different events. The events are required to realize the desired behavior of the dynamic element. Possible events are the appearance of a dynamic element, the change of the velocity, or a lane change. Every event has also one or more start conditions. The maneuver scheduler uses these conditions to start the events to execute the maneuver successfully. The events can trigger one or more actions to fulfill the event.

The executer can execute the actions on different levels. If the simulation offers the direct execution of an action, the executer will use this option. In Fig. 7.8, this is demonstrated through the block *simulation control*. If it is possible to command a lane change to a dynamic vehicle through an interface of the simulation software, the executer uses this interface to command the lane change for instance.

If such an interface does not exist, the executer has the option to command the action to dynamic elements on a lower level. This could be the driver commands to control the dynamic vehicle. For the example of the lane change, this means that the maneuver is more configurable and a trajectory can be defined for the lane changing vehicle. Thereby, the distance between the dynamic element and the ego-vehicle during the lane change can be chosen. Thereby, different kinds of lane change maneuvers for the dynamic vehicles can be simulated.

If the action cannot be executed with driver commands, the executer has also the opportunity to realize the action by setting the state of the dynamic element in every frame of the simulation. Thereby, it is possible to execute actions, which are not covered by the vehicle dynamics model of the simulation environment, a spinning vehicle for example.

With this control of the dynamic elements, the unified scenario generation has the possibility to generate static as well as dynamic scenarios with different elements. Furthermore, it is possible to generate reproducible scenarios from the perspective of the ego-vehicle.

7.4.4 Level 4: Environmental Conditions

On the first three levels of the model, the road, possible adaptions to the road, as well as the behavior of dynamic elements are described. On the fourth level of the unified scenario generation, it is possible to vary the environmental conditions, which are the daytime and the weather in the scenario for example. Possible states for the daytime are dawn, day, dusk, and night. Therefore, advanced driver assistance systems can be tested under different lighting conditions, which is mainly relevant for vision-based systems. Another option is to vary the weather conditions. The following states are possible: *sunny*, *cloudy*, *rainy*, and *snowy*. These weather states are also influencing the lighting conditions and the road conditions.

7.4.5 *Summary of the 4-Level Model of the Scenario Generation for Advanced Driver Assistance Systems*

The last section described the structure of the unified scenario generation based on the 4-level model. On the four levels, the road geometry and topology, possible adaption of the road for special situations, the dynamic elements, and environmental conditions can be defined. Figure 7.9 illustrates an example of a generated scenario based on the 4-level model for a constriction assist. The basic road is described by a straight line and a following curve. The road has four driving lanes, each with a width of 3.5 m. The second level describes an adaption by a roadwork zone. Two dynamic vehicles are defined, which are driving in front of the ego-vehicle (white vehicle in Fig. 7.9) through the road work zone. On the level of the environmental conditions, the weather is defined as rainy and the daytime as day.

On every level an amount of parameters and their values can be varied. A single scenario with five impact parameters and six discrete values for each parameter generates $6^5 = 7776$ test cases. By adding new parameters, the number of test cases increases exponentially with each parameter added and additionally with each new discrete parameter value. Thus, it is not possible to execute all test cases, if many new parameters are added due to time and economic limitations, especially when all parameter combinations have to be tested. This is the reason why new methods have to be found to reduce the number of test cases. This problem is already known and addressed in the research area of software-testing. One option is the usage of combinatorial algorithms. These algorithms reduce the number of test cases on the basis of combinatorial considerations. Kuhn et al. (2004) show in a case study that a large number of faults can be found with combinatorial testing. In this case

Fig. 7.9 Example of a generated scenario on the basis of the 4-level model for the constriction assist

study, different kinds of systems are tested. Thereby, all systems have to be tested by a large number of input parameters and parameter values. The testing of advanced driver assistance systems has similar requirements to the test method. For this reason, the efficient, systematic test case generation also uses combinatorial algorithms to reduce the number of test cases. The single steps for reducing the number of test cases are explained in the following part.

7.5 Systematic Test Case Generation

Every level of the unified scenario generation uses the concept of the systematic test case generation. The generation consists of three steps: equivalence class generation, boundary value analysis, and combinatorial test case generation. These test processes are selected out of the pool of black-box test methods. Black-box tests do not use the internal structure or the source code of the test object for the test process. The test is generated only on the basis of specifications, input and output parameters, or interface descriptions. The test results are assessed on the input and output values. The results, which are generated by the given input parameters, are compared with the expected values and checked for correctness (Liggesmeyer 2009).

7.5.1 Equivalence Class Generation

The equivalence class generation is a black-box test method. The aim of this method is to generate a high test depth with a limited number of test cases. The main idea of the equivalence classes is that all representatives of one class will evoke the same effects to the test object. Therefore, only one representative of each equivalence class has to be tested to get the results for each class.

A major challenge of the generation of the equivalence classes is the discretization of value-continuous parameters and the subsequent collection of discretized parameter values to equivalence classes. Finding an effective and efficient method for this process is still an open research question, because this is also a domain specific question.

Due to the discretization, currently a lot of values of value-continuous parameters are not tested. There is a chance that a value, which is between two discretization steps, generates a failure in the system. If this parameter value occurs during operation in the real world, the system will fail in this situation. Currently, this failure cannot be detected through this test concept, if the discretization steps are inaccurate.

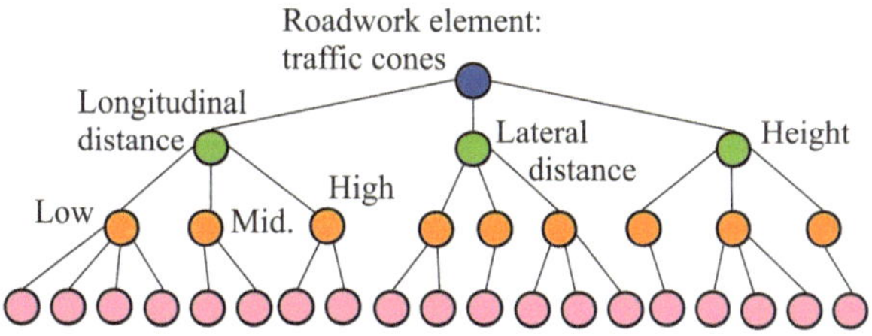

Fig. 7.10 Equivalence class for the roadwork element: traffic cone

To avoid inaccurate discretization steps, the steps can be deduced for instance on the basis of the sensor resolution or technical setup of the driving function. Additionally, specifications, standards, guidelines, and the knowledge of experts can be used to generate equivalence classes.

Figure 7.10 shows an equivalence class for traffic cones for road works. These can be subdivided into the longitudinal and lateral distance between the cones and the height of the cones. In these classes, the continuous values are reduced to discrete value ranges, for example from 0.7 to 3 m (Low), 3 to 10 m (Mid.), and 10 to 20 m (High) for the longitudinal distance.

The equivalence classes can be distinguished in valid and invalid classes. A valid class contains only values, which are in the specification. Invalid classes contain also values, which are out of the range of the specification. The test of advanced driver assistance systems should use valid and invalid equivalence classes to test the system for robustness and the behavior at the limits of the specification.

In the systematic test case generation, the equivalence class generation is used to discretize the value-continuous impact parameters into discrete values and to test the system with values in and out of the range of the specification.

7.5.2 Boundary Value Analysis

The boundary value analysis describes an optional expansion to the equivalence class generation. The difference between equivalence class generation and boundary value analysis is the strategy to find a suitable representative in the class of possible parameter values. Whereas in equivalence class generation an arbitrary representative is chosen for the representative test case, the boundary value analysis uses the limit values of a class to find a representative test case. Figure 7.11 illustrates the boundary value analysis for the example of the traffic cones in road work zones. Thereby, it is possible to select a representative, which is more fault-sensitive as the other representatives (Liggesmeyer 2009). The crosses mark classes are not selected by the boundary analysis.

The boundary value analysis should also be used in the systematic test case generation. Thus, the relevant values of the impact parameters for the test cases have to be selected. With the boundary value analysis, an efficient preliminary

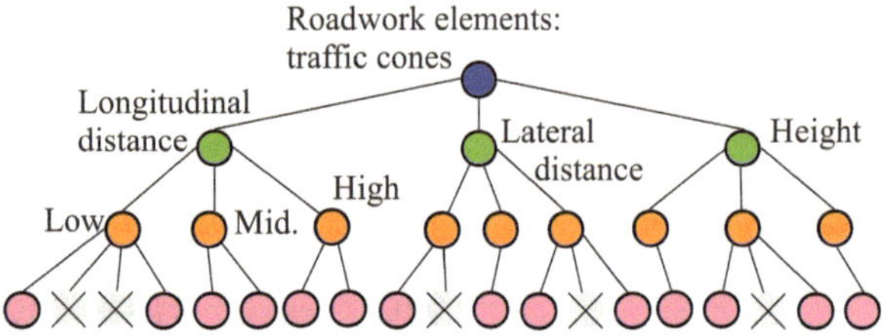

Fig. 7.11 Boundary value analysis for the roadwork element: traffic cone

selection of the parameter values can be created, based on requirements, standards, guidelines, and experts' knowledge.

However, the combinations of remaining parameter values also generate a large number of possible test cases. According to time restrictions for the test process, all test cases cannot be executed. As a consequence, the number of scenarios has to be further reduced. The efficient test method reduces the number of test cases through combinatorial methods, which are explained in the following section.

7.5.3 Combinatorial Test Case Generation

In the last two sections, methods to reduce the number of possible test parameter values were presented. With these two methods, it is still not possible to test all combinations of the parameter values, because the number of resulting test cases is too high. To solve this problem, the efficient test method will use the principle of combinatorial test methods to generate a test suite with a reduced number of test cases. The aim of the combinatorial test methods is the generation of non-redundant test cases by varying the given parameter values. Thus, the number of test cases in the test suite can be reduced by well-known test case coverage. The combinatorial test suite coverage and some algorithms to generate combinatorial test suites are presented in the following part.

7.5.4 Test Coverage for Combinatorial Test Case Generation

The combinatorial test case generation can generate different test suites with different coverage criteria. The weakest coverage criterion is the *each-used* coverage. This criterion is reached, if every value of a parameter is presented at least in one single test case. This coverage criterion can be generated with a relatively small number of test cases in the test suite. However, the results of the test execution with this test suite cannot present a representative test of the test object, because a lot of faults occur by special combinations of two or more parameter values (Kuhn et al. 2004). These combinations will not be tested by the *each-used* coverage and the faults, which occurred by a combination of two or more parameter values, are not discovered.

A test criterion with a higher coverage is the *pair-wise* test coverage. This criterion is satisfied, if every pair of values of the parameters is presented in at least one test case. On the one hand, a test suite with more test cases is necessary to satisfy this test coverage. On the other hand, more faults can be found with this test coverage, because the interactions of the parameter values are tested. Every pair combination of parameter values is presented at least in one single test case. The test coverage can be expanded, if a *t-wise* (combination of t parameter values) coverage is chosen. In this case *t-wise* combinations are used instead of *pair-wise* combinations, for example a combination of triples, if a *3-wise* coverage is chosen.

The highest test coverage is *N-wise*. To reach this coverage, all possible combinations of the parameter values have to be tested in the test suite.

The following example will demonstrate the evolution of the number of test cases with the increasing test coverage. The scenario has three input parameters (A, B, and C), each with four values (1, 2, 3, or 4).

To reach the each-used coverage, only four test cases are required (see Table 7.3).

For a *pair-wise* coverage, 16 test cases are necessary (see Table 7.4).

Table 7.3 Test suite with an each-used test coverage

	Parameter	A	B	C
Each-used	Test case 1	A1	B1	C1
	Test case 2	A2	B2	C2
	Test case 3	A3	B3	C3
	Test case 4	A4	B4	C4

Table 7.4 Test suite with 16 test cases for a pair-wise test coverage

	Parameter	A	B	C
Pair-wise	Test case 1	A1	B1	C1
	Test case 2	A1	B2	C2
	Test case 3	A1	B3	C3
	Test case 4	A1	B4	C4
	Test case 5	A2	B1	C2
	Test case 6	A2	B2	C3
	Test case 7	A2	B3	C4
	Test case 8	A2	B4	C1
	Test case 9	A3	B1	C3
	Test case 10	A3	B2	C4
	Test case 11	A3	B3	C1
	Test case 12	A3	B4	C2
	Test case 13	A4	B1	C4
	Test case 14	A4	B2	C1
	Test case 15	A4	B3	C2
	Test case 16	A4	B4	C3

According to the given example with three parameters, a *3-wise* test coverage is equivalent to an *N-wise* coverage, which requires 64 test cases. This small example shows, how the combinatorial test case generation can reduce the number of test cases in a test suite with defined test coverage.

7.5.5 Algorithms for Combinatorial Test Case Generation

In literature, some algorithms are given to generate test suites with defined combinatorial test coverage. The algorithms can be divided into deterministic and non-deterministic algorithms (Grindal et al. 2005; Nie and Leung 2011).

Deterministic algorithms generate the same test suite for the given input parameter values. Furthermore, the deterministic algorithms can be divided into algorithms, which generate the test suite instantly or iteratively. Examples for instant generating algorithms are orthogonal arrays (OA) or covering arrays (CA). These algorithms can generate a combinatorial test suite very fast, but they cannot be used for every test scenario, because the orthogonal and covering arrays are only available for special test setups with a defined number of parameters and values for each parameter (Grindal et al. 2005). Alternative approaches are iterative working algorithms. These algorithms can be used on every test setup, because the algorithms are independent from the test setup. The test suite can be generated on the basis of parameters or test cases.

The IPOG algorithm (In Parameter Order General) is one example for an iterative parameter based algorithm (Lei et al. 2008). *IPOG* generates a test suite oriented on the parameters and their values, which can satisfy a *t-wise* coverage. An advantage of this kind of algorithms is that new parameters can be added to a preexisting test suite without a regeneration of the whole test suite. Additionally, conditions between the parameters can be defined to generate only test suites with representative test cases.

One algorithm for a test case based generation is *base choice*. The *base choice* algorithm generates a base test with the most important value of each parameter. In the next steps new test cases are added to the test suite by varying the value of one parameter.

The non-deterministic algorithms generate always different test suites for one test scenario. Here, the algorithms can be divided into the groups of heuristic, artificial life, and random algorithms. Examples for heuristics are the algorithms *Automatic Efficient Test Generator (AETG)* (Cohen et al. 1997) or *Simulated Annealing* (Cohen et al. 2003). Both algorithms can satisfy the test coverage of *pair-wise* tests. *Genetic algorithms* and *Ant colony algorithms* are artificial life algorithms (Shiba et al. 2004). These algorithms can also satisfy the *pair-wise* test coverage.

Based on the amount of impact parameters and their values, which affect advanced driver assistance systems, the system test is a difficult task. Such systems should be tested for reasons of safety and ruggedness by an algorithm, which generates a test suite with an *N-wise* test coverage. However, it is not possible by reasonable expenditure. For this reason, a strategy with a *pair-wise* or *t-wise* coverage should be chosen to generate a test suite. In a lot of cases, a fault is based on the combination of two parameter values (Kuhn et al. 2004). Thereby, the algorithms AETG, OA, CA, and IPOG are suitable to test advanced driver assistance systems. In the case of the different test scenarios with a varying number of parameters and their values, the algorithm should be flexible for changing test setups. Thus, the best choice for a combinatorial test case generation is the IPOG algorithm, because the algorithm is deterministic and the generation of test cases is based on the impact parameters. The generated test suite can be extended easily with new parameter values or parameters, which will be identified at a later moment.

7.6 The Case Study: Constriction Assist

Parts of the efficient, systematic test method and the efficient test case generation will be demonstrated by the case study of the constriction assist.

The constriction assist has been developed in the research project UR:BAN. The assistance system supports the driver by the lateral guidance in the situations of narrowed lanes, e.g. road works, or "when passing vehicle platoons in neighboring lanes, fixed obstacles, or parking cars. A warning is given if the constriction is too narrow to pass through". The system is developed for urban traffic (Scholl 2015).

The driver assistance system will be tested in a roadwork scenario. One typical scenario in road works, which are derived from the standards and guidelines, is the drive through a chicane. Figure 7.12 illustrates one possible configuration of such a chicane. The standards and guidelines provide a standardized range of ratios between the length and the lateral offset of the chicane. A ratio of 1:10 means that the length of the chicane has to be 35 m, if the lateral offset is 3.5 m. Furthermore, the lateral offset, the type of the roadwork elements, the distance of the roadwork elements to each other, and the lane width within the chicane have to be defined to generate a scenario in a road works.

These parameters are identified through an analysis of the standards and guidelines for road works in Germany (FGSV 2009). However, the standards and guidelines provide no concrete discretization steps for these value-continuous

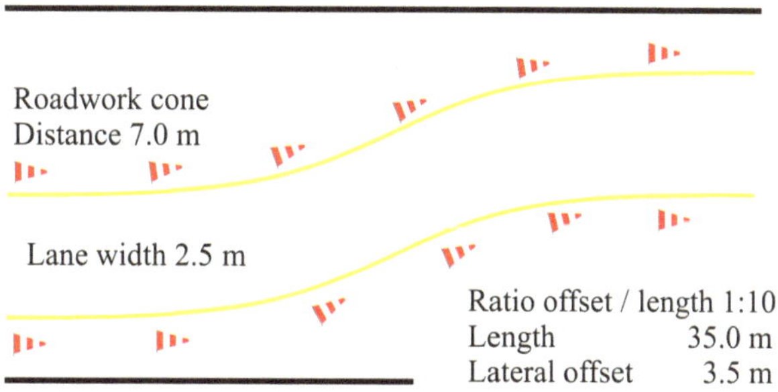

Fig. 7.12 A roadwork scenario for the constriction assist

parameters for a test case generation. Only minimal and maximal allowed values are provided in the standards and guidelines for instance the minimal allowed lane width in road works is 2.5 m. The values have to be discretized for the test case generation with the method of the efficient, systematic test case generation.

First, equivalence classes are generated for the impact parameters. The minimal and maximal allowed values in the standards and guidelines are also the minimal and maximal boundaries in the equivalence classes.

For the lane width, the minimal allowed value is 2.5 m. Therefore, this value is also the lower boundary of the equivalence class. The upper boundary is 3.75 m, because this is the maximal possible lane width according to standards in Germany. Between the minimal and maximal value, the values are linear interpolated by 0.25 m within the class. With these steps, all typical lane widths in road works and cross sections are covered (FGSV 1996, 2009).

The generation of an equivalence class for roadwork elements is trivial, because only three kinds of elements in road works exist in Germany (FGSV 2009). These are traffic beacons, cones, and concrete barriers.

The lateral offset of the chicane is based on the typical lane width of 3.5 m in Germany. It will be assumed, that the lateral offset of the chicane is a multiple of the typical lane width in Germany and the lane is always swiveled by complete standard lane widths. Intermediate values are not tested with a remaining risk that a failure occurs for these values.

The distance of roadwork elements are also derived from the standards and guidelines. Here, the provided distance lies between 5.0 and 10.0 m in urban environments (FGSV 2009). To test also road works, which are not correctly constructed to the standards, the distance range is extended up to 13.0 m. The range between 5.0 and 13.0 m is again linear interpolated by the distance of 2.0 m in the first iteration of the test method.

The ratio between the lateral offset and the length of the chicane is also provided in the standards and guidelines (FGSV 2009). Here, the standards provide a ratio between 1:10 and 1:20 for the chicane. To test the system also outside of the standards, the values of 1:5, 1:7, and 1:25 are introduced. Additionally, the value 1:15 is introduced to interpolate between the provided values. With these values, invalid and valid equivalence classes can be generated to test the system boundaries.

Table 7.5 shows the identified parameter values for the scenario.

Table 7.5 Identified parameter values for the scenario

Impact Parameter	Parameter values					
Lateral offset [m]	3.5	7.0	10.5	14.0		
Ratio lateral offset/length	1:5	1:7	1:10	1:15	1:20	1:25
Roadwork elements	Traffic beacons	Cone	Concrete barrier			
Distance roadwork elements [m]	5	7	9	11	13	
Lane width [m]	2.50	2.75	3.00	3.25	3.5	3.75

On the basis of the identified parameter values, a test suite is generated with the combinatorial algorithm IPOG (Lei et al. 2008). A *pair-wise* test coverage is chosen to analyze the main effects of the parameter values. With this configuration, the algorithm generates 36 test cases. Table 7.6 shows all these test cases

Table 7.6 Generated test suite on the basis of the identified parameter values

Test case	Ratio of lateral offset and length	Lateral offset [m]	Roadwork elements	Distance of roadwork elements [m]	Lane width [m]
1	1:5	3.5	Traffic beacons	5	2.50
2	1:5	7.0	Cone	7	2.75
3	1:5	10.5	Concrete barrier	9	3.00
4	1:5	14.0	Traffic beacons	11	3.25
5	1:5	3.5	Cone	13	3.50
6	1:5	7.0	Concrete barrier	5	3.75
7	1:7	10.5	Traffic beacons	7	2.50
8	1:7	14.0	Cone	9	2.75
9	1:7	3.5	Concrete barrier	11	3.00
10	1:7	7.0	Traffic beacons	13	3.25
11	1:7	10.5	Cone	5	3.50
12	1:7	14.0	Concrete barrier	7	3.75
13	1:10	7.0	Traffic beacons	9	2.50
14	1:10	10.5	Cone	11	2.75
15	1:10	14.0	Concrete barrier	13	3.00
16	1:10	3.5	Cone	5	3.25
17	1:10	3.5	Traffic beacons	7	3.50
18	1:10	3.5	Traffic beacons	9	3.75
19	1:15	7.0	Concrete barrier	11	2.50
20	1:15	10.5	Traffic beacons	13	2.75
21	1:15	14.0	Cone	5	3.00
22	1:15	10.5	Concrete barrier	7	3.25
23	1:15	7.0	Concrete barrier	9	3.50
24	1:15	3.5	Cone	11	3.75
25	1:20	14.0	Cone	13	2.50
26	1:20	3.5	Concrete barrier	5	2.75
27	1:20	7.0	Traffic beacons	7	3.00
28	1:20	10.5	Concrete barrier	9	3.25
29	1:20	14.0	Traffic beacons	11	3.50
30	1:20	10.5	Cone	13	3.75
31	1:25	7.0	Traffic beacons	5	2.50
32	1:25	14.0	Cone	7	2.75
33	1:25	3.5	Concrete barrier	9	3.00
34	1:25	10.5	Cone	11	3.25
35	1:25	10.5	Cone	13	3.50
36	1:25	14.0	Traffic beacon	13	3.75

in the test suite. Every pair of parameter values is tested in at least one single test case.

The test suite is created with the aim to generate efficient test cases. The efficient test cases have to be non-redundant, representative, unified, and reproducible.

Due to the analysis of the standards and guidelines and the identified parameters, the test cases are representative for the constriction assist. There is a risk that parameters exist apart from the identified parameters, which are also representative for the system, but not coverable through the standards and guidelines. The test suite is currently lacking these test cases. To identify these test cases, additional sources next to the standards and guidelines, like experts, should be consulted in the step of the analysis.

Due to a combinatorial test case generation and a *pair-wise* coverage, every pair of parameter values is presented by one single test. With this test suite, all test cases are non-redundant, because no scenario is tested twice. Due to the *pair-wise* coverage, faults can be detected, which are depending on two values. Faults depending on three or more parameter values cannot be detected with this test suite. Therefore, a *3-wise* or *t-wise* coverage would have to be generated.

The test cases are based on the unified 4-level model for scenarios. On the first layer, a straight line is defined as the basic road. The roadwork is defined on the second level of the model. For testing the roadwork scenario on a new basic road, only the first layer has to be modified. Thus, new test cases can be generated flexibly by varying the single levels of the 4-level model.

Due to the static scenario without dynamic objects, the test cases are reproducible, because no non-deterministic element is part of the scenario. The reproducibility of the test cases with dynamic objects cannot be demonstrated in this case study. In Symkenberg (2015), the reproducibility of scenarios with dynamic elements is demonstrated in different scenarios with the presented control of the dynamic elements.

7.7 Summary and Outlook

This chapter proposes a systematic method for the test case generation for advanced driver assistance systems in virtual environments.

The test case generation is one of four steps in the efficient, systematic test method for advanced driver assistance systems. The four steps of the method are: (1) analysis of the system, (2) systematic test case generation, (3) test case execution, and (4) test case evaluation. The analysis of the system aims at identifying parameters, which have a significant impact on the system under test. Specifications or standards and guidelines can be used to identify such significant impact parameters.

The test case generation is based on a 4-level model to describe scenarios on different levels. On each level, the principle of the systematic test case generation is used to generate test cases from the scenario. Therefore, methods out of the black-box testing and combinatorial testing are used to reduce the number of test cases.

The method of the test case generation aims to generate efficient and systematic test cases. To reach this, the method has to generate test cases, which are non-redundant, representative, unified, and reproducible. Through the analysis of the system, representative test cases are generated. With the 4-level model, unified scenarios can be defined on different levels. Due to the control of the dynamic elements, the test cases are reproducible, also when non-deterministic elements are part of the scenario. The combinatorial test case generation ensures that the test cases are non-redundant.

A deficit of this test method is a strong focus on the standards and guidelines, because scenarios, which exist in real world and outside of the standards, are not tested. Additionally, there is a deficit, if no standards and guidelines are available for the country or the environment, where the system under test has to operate.

In future research, more methods and algorithms will be implemented to reduce the number of necessary test cases for a release. Additionally, the usage of the combinatorial test case generation will be determined for scenarios with dynamic elements. Furthermore, the integration of expert knowledge for the identification of new test parameters will be analyzed as an alternative to the standards and guidelines.

Acknowledgments Special thanks to Kathrin Symkenberg for her support by the generation of dynamic scenarios in the efficient test case generation.

References

Ammann, P., Jeff, O.: Introduction to Software Testing. Cambridge University Press, New York (2008)

Bock, T.: Vehicle in the loop – Test- und Simulationsumgebung für Fahrerassistenzsysteme. PhD Dissertation, Technische Universität München, INITUM (2008)

Bock, T., Maurer, M., Färber, G.: Validation of the vehicle in the loop (VIL) – a milestone for the simulation of driver assistance systems. In: Proceedings of the Intelligent Vehicles Symposium (IV), Istanbul, Turkey, pp. 612–617 (2007)

Cohen, D., Dalal, S.R., Fredman, M.L., Patton, G.C.: The AETG system: an approach to testing based on combinatorial design. Trans. Softw. Eng. **23**(7), 437–444 (1997)

Cohen, M.B., Gibbons, P.B., Mugridge, W.B., Colbourn, C.J.: Constructing test suites for interaction testing. In: Proceedings of International Conference on Software Engineering, Portland, OR, pp. 38–48 (2003)

Eltaher, A.: Human-like test systems: a cognitive-oriented approach applied to infotainment devices. PhD Dissertation, Technische Universität Braunschweig, Shaker Verlag (2013)

FGSV: Richtlinien für die Markierung von Straßen, Forschungsgesellschaft für Strassen- und Verkehrswesen. Arbeitsgruppe, FGSV Verlag, Köln (1980)

FGSV: Richtlinien für die Anlage von Straßen Teil Linienführung RAS-L, Forschungsgesellschaft für Straßen- und Verkehrswesen. Arbeitsgruppe, FGSV Verlag, Köln (1995)

FGSV: Richtlinien für die Anlage von Straßen Teil Linienführung RAS-Q 96, Forschungsgesellschaft für Straßen- und Verkehrswesen. Arbeitsgruppe, FGSV Verlag, Köln (1996)

FGSV: Richtlinien für die Anlage von Stadtstraßen, Forschungsgesellschaft für Straßen- und Verkehrswesen. Arbeitsgruppe, FGSV Verlag, Köln (2006)
FGSV: Richtlinien für die Anlage von Autobahnen, Forschungsgesellschaft für Straßen- und Verkehrswesen. Arbeitsgruppe, FGSV Verlag, Köln (2008)
FGSV: Richtlinien für die Sicherung von Arbeitsstellen an Straßen, Forschungsgesellschaft für Straßen- und Verkehrswesen. Arbeitsgruppe, FGSV Verlag, Köln (2009)
Geyer, S., Baltzer, M., Franz, B., Hakuli, S., Kauer, M., Kienle, M., Meier, S., Weißgerber, T., Bengler, K., Bruder, R., Flemisch, F., Winner, H.: Concept and development of a unified ontology for generating test and use-case catalogues for assisted and automated vehicle guidance. IET Intell. Transp. Syst. **8**(3), 183–189 (2014)
Gietelink, O.J., Ploeg, J., De Schutter, B., Verhaegen, M.: VEHIL: test facility for fault management testing of advanced driver assistance systems. In: Proceedings of the 10th ITS World Congress, Salerno, Italy, pp. 397–402 (2003)
Grindal, M., Offutt, J., Andler, S.F.: Combination testing strategies: a survey. Softw. Test. Verif. Reliab. **15**(3), 167–199 (2005)
Hendriks, F., Pelders, R., Tideman, M.: Future testing of active safety systems. SAE Int. J. Passeng. Cars - Electron. Electr. Syst. **3**(2), 170–175 (2010)
Hilf, K.D., Matheis, I., Mauss, J., Rauh, J.: Automated simulation of scenarios to guide the development of a crosswind stabilization function. In: 6th IFAC Symposium Advances in Automotive Control, Munich, Germany, pp. 751–756 (2010)
ISO 22179: Intelligent Transport Systems – Full Speed Range Adaptive Cruise Control (FSRA) Systems – Performance Requirements and Test Procedures. ISO, Geneva (2009)
ISO 15622: Intelligent Transport Systems – Adaptive Cruise Control Systems – Performance Requirements and Test Procedures. ISO, Geneva (2010)
ISO 26262: Road Vehicles – Functional Safety. ISO, Geneva (2011)
ISO 11270: Intelligent Transport Systems – Lane Keeping Assistance Systems (LKAS) – Performance Requirements and Test Procedures. ISO, Geneva (2014)
Kuhn, D.R., Wallace, D.R., Gallo Jr., A.M.: Software fault interactions and implications for software testing. IEEE Trans. Softw. Eng. **30**(6), 418–421 (2004)
Lei, Y., Kacker, R., Kuhn, D.R., Okun, V., Lawrence, J.: IPOG/IPOG-D: efficient test generation for multi-way combinatorial testing. Softw. Test. Verif. Reliab. **18**(3), 125–148 (2008)
Liggesmeyer, P.: Software-Qualität. Spektrum-Verlag, Heidelberg (2009)
Lindlar, F.: Modellbasierter evolutionärer Funktionstest. PhD dissertation, Technische Universität Berlin (2012)
Nie, C., Leung, H.: A survey of combinatorial testing. ACM Comput. Surv. **43**(2), 1–29 (2011)
Rook, P.: Controlling software projects. Softw. Eng. J. **1**(1), S.7–S.16 (1986)
Schmidt, F.: Funktionale Absicherung kamerabasierter Aktiver Fahrerassistenzsysteme durch Hardware-in-the-Loop-Tests. PhD Dissertation, Technische Universität Kaiserslautern (2012)
Scholl, W.: URBAN Homepage, Urbaner Raum: Benutzergerechte Assistenzsysteme und Netzmanagement (2015). Accessed 30 Mar 2015
Schuldt, F., Menzel, T., Maurer, M.: Eine Methode für die Zuordnung von Testfällen für automatisierte Fahrfunktionen auf X-in-the-Loop Simulationen im modularen virtuellen Testbaukasten. In: Workshop Fahrerassistenzsysteme, Walting, Germany, pp. 171–182 (2015)
Shiba, T., Tsuchiya, T., Kikuno, T. Using artificial life techniques to generate test cases for combinatorial testing. In: Proceedings of the Computer Software and Applications Conference (COMPSAC), Hong Kong, China, pp. 72–77 (2004)
Siebertz, K., van Bebber, D., Hochkirchen, T.: Statistische Versuchsplanung. Springer Science & Business Media, Berlin (2010)
Sommerville, I.: Software Engineering, 8th edn. Addison-Wesley, Harlow (2007)
Symkenberg, K.: Entwicklung und Implementierung eines Frameworks zur systematischen Gestaltung und Variation von dynamischen Szenarien zur Testung von Fahrerassistenzsystemen. Masterthesis, Universität Hannover (2015)

Taguchi, G., Chowdhury, S., Wu, Y.: Introduction to the quality loss function. In: Taguchi, G., Chowdhury, S., Wu, Y. (eds.) Taguchi's Quality Engineering Handbook. Wiley, Hoboken, NJ (2007)

Ulbrich, S., Menzel, T., Reschka, A., Schuldt, F., Maurer, M.: (2015) Defining and Substantiating the terms scene, situation and scenario for automated driving. In: Proceedings of Intelligent Transportation Systems (ITSC), Las Palmas, Spain (2015)

Verhoeff, L., Verburg, D.J., Lupker, H.A., Kusters, L.J.J.: VEHIL: a full-scale test methodology for intelligent transport systems, vehicles and subsystems. In: Proceedings of the IEEE Intelligent Vehicles Symposium (IV), Dearborn, USA, pp. 369–375 (2000)

Chapter 8
Validation and Introduction of Automated Driving

Hermann Winner, Walther Wachenfeld, and Phillip Junietz

Abstract With the introduction of automated driving without driver supervision, the automotive industry breaks new ground not just in functionality, but also in terms of validation. Even the most extensive road tests cannot statistically prove the safety of an automated vehicle. This makes the use of alternative validation techniques necessary. In principle, these techniques are already known but in order to apply them many fundamentals must still be determined, especially the base amount of required field data. Thus, methods for obtaining this data from road testing and field applications gain a high importance for future safety certification as a basis for the approval of vehicle automation systems.

Keywords Approval • Automated driving • Safety • Validation methods

8.1 The Challenge: Validation of Automated Driving

Automated driving without supervision, also known as automated driving level three or higher on the scales of BASt (Gasser et al. 2012; Verband der Automobilindustrie 2015; NHTSA 2013; SAE International Standard J3016 2014), poses a new challenge to the validation of automated vehicles. Certain aspects of the validation of automation levels 1 and 2 (called AD2−) can be reused but are considerably different than those of future automated driving functions (called AD3+).

The main difference between AD3+ and continuously assisting driver assistance systems (level 1), namely longitudinal (e.g. Adaptive Cruise Control), lateral (e.g. Lane Keeping Assist), and combined systems (level 2), is that assistance systems abstain from rough and abrupt interventions (Winner et al. 2016a). Due to the predictability and low intensity of (non-)reaction, the driver is always able to override AD2− system commands. This is complemented by the prompt for the driver to take control when the system limits are reached. Thus, with overtaking

H. Winner (✉) • W. Wachenfeld • P. Junietz
Fachgebiet Fahrzeugtechnik, Technische Universität Darmstadt, Otto-Berndt-Str. 2, 64287 Darmstadt, Germany
e-mail: winner@fzd.tu-darmstadt.de

H. Winner et al. (eds.), *Automotive Systems Engineering II*,
DOI 10.1007/978-3-319-61607-0_8

times on the order of 1s, dangerous situations due to system limits can be prevented. Some intervening systems, namely automated emergency braking systems, do intervene in a rough way, but only if the driver ignores previous distinct system warnings, making the use of rough interventions a rare occurrence. In most cases the driver will have reacted in an appropriate way in response to the previous warnings, for instance by emergency braking. Besides the precondition of having passed all previous warnings without the driver reacting, a high confidence in signal processing must be reached, guaranteeing that a detected situation is truly dangerous. If a supporting system is unsure whether or not a situation is actually dangerous, it abstains from sharply intervening. Systems of both categories are designed for controllability, even if this leads to a reduction of their functionality and therefore their benefit. Accordingly, evidence of their controllability is the primary focus for their validation, see ISO 26262 and Code of Practice (Schwarz 2006). Ensuring the functional quality "only" serves the adherence to the greatest benefit for the customer, ensures that intervening systems pass the functionality tests of customer organizations (e.g. EURO-NCAP) and that systems designed for trucks pass the licensing regulations. Extended road testing for data collection is conducted in order to define threshold values for intervening systems. With the acquired data, algorithms for triggering interventions are optimized so as to in first place minimize the frequency and impact of incorrectly triggered (false positive) interventions and in second place maximize the amount of legitimate (true positive) interventions. With assistance systems a suppression of the emergency system is considered a safe state, as there is no aggravation compared to the previous state without the assistance system. This, as well as only acting within a certain comfort zone, is not an option for automated driving without driver supervision. Unsupervised automated driving demands that a system exploits the boundaries of driving physics as needed. Nevertheless, questions regarding the controllability of the transition between automated and manual driving still remain the subject of deliberation. The overarching question regarding validation concerns how automated driving can be proven to a safety level that is generally accepted in comparison with the current safety of human drivers.

8.2 Safety References

A state is considered safe if the current risk is below an accepted risk, where risk is the combination of the probability and severity of damage. If the damage can be quantified across multiple damage classes, then the risk is generally defined as the product of the relative frequency of occurrence with the absolute damage. When the damage cannot be assessed as a scalar, only the (relative) frequency of occurrence in the respective damage class can be specified (Baum et al. 2011). On the one hand, the insurance business does provide monetary amounts of loss for a variety of damage categories that are used for the economic consideration of damage caused by accidents Gasser et al. (2012), but on the other hand, due to the special meaning of the conservation of life and physical inviolability (see Article 2 (2), constitution of Germany), comparing risk of human lives with material costs is morally difficult to justify.

In 1980, the micromort unit (μM) was introduced in order to quantify the absolute risk of death, where one micromort is a one-in-a-million chance of fatality (Schwing and Albers 2013).

Beside the differentiation into categories of damage and loss, a reference must be found. On the one hand, a time-wise consideration can be made, considering the period of time that a person is exposed to road traffic (exposure time). This kind of risk can be compared to other time-related measures, for example the minimum endogenous mortality (MEM), defined to be 200 μM/a (μM per annum) by the EN50126 standard. MEM is the natural death rate: the total death rate without deaths from accidents or congenital malformations. While the MEM defined in EN50126 uses numbers from 1974 the MEM calculated from today's mortality tables amounts to about a third of the value specified in the standard. This amounts to $2.3 \cdot 10^{-2}$ μM/h in the standard and $0.8 \cdot 10^{-2}$ μM/h if calculated using today's minimum endogenous mortality. Values of 1.1 and 1.3 μM/h can be derived from average mortality and life expectancy (statista.de 2013).

However, traffic exposure time is neither known, nor is it anyone's objective to spend excessive time in traffic. Thus, it is sensible to refer to transportation. This can be measured according to passenger service work, measured in passenger kilometers (pkm).

Table 8.1 shows values for several modes of transport ranging from 16 pkm/μM (Motorcycle) to over 300,000 pkm/μM (Aviation). Motor vehicle drivers experience a value of 341 pkm/μM. However, as vehicle drivers are better protected in an accident than potentially affected pedestrians or cyclists, only one part of the death risk is addressed. The number of deaths per distance can be used as a second measure to compare risk. Taking the data from the 2013 accident statistics for Germany, we arrive at a value of 4.6 μM/1000 km (1.9 μM/1000 km on Autobahn, see Table 8.2) (Statistisches Bundesamt 2013a). This reference seems to be suitable as a measure to compare risks if automated vehicles are deployed in the same way as those driven by humans. An automated Autobahn-chauffeur restricted to use on Autobahn must therefore have its risk compared to the number of deaths per Autobahn kilometer. This consideration is more difficult for automated vehicles that provide new mobility services because there is no existing reference data. Searching for safety target values to validate, the number of fatalities caused by the deployment of the new technology seems sufficient. However, this value should not be used to assess safety because the number of fatalities caused by accidents depends

Table 8.1 Overview of passenger kilometers per death for several modes of transport (calculation from Vorndran 2010; Statistisches Bundesamt (Destatis) 2015)

Mode of transport	Passenger kilometers per death
Motorcycles	16 pkm/μM
Motor vehicles	341 pkm/μM
Local public transport	2.7 pkm/μM
Aviation	>300,000 pkm/μM

Table 8.2 German 2013 accident statistics (Statistisches Bundesamt 2013a)

Category	Number of casualties	Number of fatalities
Due to traffic accidents	377,481	3339
Per 1000 motor vehicles	6.9	0.1
Per billion motor vehicle km	520	4.6
Portion on Autobahn	132	1.9
Per 1000 accidents with damage to persons	1297	11.5
On Autobahn	1606	23.2
Per million habitants	4681	41
Fatalities in traffic accidents		
Users of passenger vehicles		1588
Users of commercial vehicles		148
Users of motorcycles		568
Users of bicycles		554
Pedestrians		557

on many irrelevant circumstances from the technical point of view, for example the number of passengers. Furthermore, there is currently no better alternative than to assume the number of fatalities per accident to be constant. As soon as new values for this purpose exist, they can be adapted by means of a conversion factor redefining the reference distance per accident. So instead of the number of additional fatalities, the number of accidents per distance (for the respective category of accident consequences) should be used. Values between two fatal accidents ranging from 10^8 to 10^9 km are typical. Accidents with damage to persons occur approximately 50 times more often (Statistisches Bundesamt 2013b).

8.3 Statistical Proof of Safety

The travel distances necessary to statistically prove the safety of an automated system can be calculated under certain assumptions using a reference for the expected distance between accidents of certain severities derived in the previous section. For a more detailed look at the derivation of our results, see Winner et al. (2015) and Wachenfeld and Winner (2016). Supposing that the testing route of automated vehicles is representatively selected and that accidents occur independently from each other with a constant expected value ($\Rightarrow$ Poisson distribution), one obtains a validation drive distance that is up to 20 times the reference value for a vehicle that is twice as safe as this respective reference (e.g. humans). Thus, the required testing distances grow to over 10^8 km (counting accidents with injuries) or 10^{10} km on Autobahn for fatal accidents. Today, assuming that one motor vehicle travels 5000 Autobahn-km per year, the latter distance of 10^{10} km required to thoroughly test an automated vehicle would require two million motor vehicles to sufficiently cover the distance in 1 year, involving nearly half of all new passenger cars annually registered in Germany! This leads to the conclusion that a statistical proof of safety of an automated vehicle cannot be provided before its market launch, but during an ongoing observation of an adequately large number of vehicles in regular traffic after release.

Then why is this statistical proof even necessary? In the past, innovations have successfully been deployed in traffic without such extensive testing. In aviation, where the reference distances are far higher, this kind of extensive testing is not even considered. Why should we break with this tradition for automated driving?

8.4 The Knowledge Gap of Automated Driving

The formerly discussed validation is marked in green in the combined illustration of Rasmussen's three-level model for targeted human activities (Rasmussen 1983) and Donges' three-level hierarchy of the driving task (Donges 1982) illustrated in Fig. 8.1. The vehicle and its behavior in lateral and longitudinal directions are tested.

Through this we do not test the skills or abilities of the future driver, but the ability of the test driver to control the vehicle in test cases with the steering wheel and pedals. Thus, the green box in Fig. 8.1 only slightly overlaps with the area of the driver.

For automated driving without driver supervision, the abilities of the driver are omitted, as well as his role as a backup driver. The driving tasks of navigation, guidance, and stabilization are adopted by the driving robot. This means that for automated driving only functionality is testable, no system controllability. On the one hand, this facilitates the test because human uncertainties and individual differences do not need to be covered. On the other hand, the possibility to extrapolate from test cases and test drivers to other applications is removed. The human aspect of the system is also omitted, and is typically replaced with skill-based, rule-based, and knowledge-based systems. When validating an automated driving

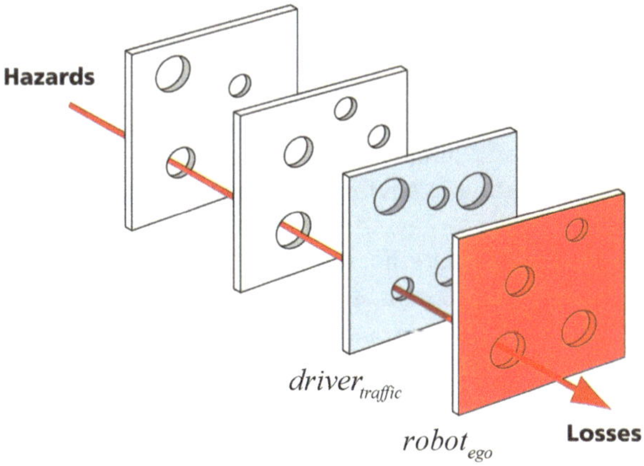

Fig. 8.1 Rasmussen's three-level model for targeted human activities and Donges' three-level hierarchy of the driving task (Wachenfeld and Winner 2016 based on Donges 1982; Rasmussen 1983)

system, the safety that has to be proven only comes from the technical system consisting of the driving robot and vehicle (yellow field in Fig. 8.1). However, human aspects are still relevant in the context of a mixed traffic, so the behavior of autonomous vehicles has to be compatible to human traffic participants.

As can be seen from Fig. 8.1, the number of tasks that must be validated is increased: The driving robot takes over diverse applications, including navigation, guidance, and stabilization. This increased task quantity will be a challenge, especially in public areas without restriction on access. Task quality has also changed. Current state-of-the-art systems merely perform under human supervision, while fully automated systems must fulfill their tasks in a way that satisfies the safety demands discussed in the beginning.

What is known about the driving task apart from qualitative models? Quantitative models, that describe car following, intersection navigation, and lane-changing already exist (Reichart 2001a; Schnieder and Schnieder 2013). However, these models do not address rare accidents, especially because these generally depend on local or temporary circumstances which do not show up in generalized statistics. Furthermore, reliability models (e.g. Reichart 2001b) only allow quantitative statements with great uncertainty and therefore only are suitable for descriptive instead of predictive applications. Hence, only records of past accidents, such as from police records or in-depth-analyses of special projects like the German GIDAS project, are left to work with.

8.5 Safety Prediction Model

For mechanical or electrical components, failure models can be found either from experience or through special lab testing. These failure models can then be used to predict how long and under which circumstances the components can meet their requirements. However, for tasks that up to now have only been fulfilled by humans, e.g. driving a vehicle, no models exist. In the end, only unwanted failure cases are recorded namely accidents. Throughout the following approach, this kind of failure is modeled in an overly simplified way.

A key element of this model is the *critical scenario* defined in the following manner:

A segment, limited either in time or distance, to which surrounding circumstances like traffic environment, intentions, and trajectories of the traffic participants relevant for the criticality are known. A state is critical when the criticality metric passes a threshold value. Such a criticality metric is initially arbitrary but relevant for the instantiation.

The number of accidents $n_{\mathrm{ac,hd}}$ caused by human drivers (index hd) is modeled as the product of the number of critical scenarios $n_{\mathrm{crit,hd}}$ experienced by the driver and the transition probability $\rho_{\mathrm{tr,hd}}$:

$$n_{\mathrm{ac,hd}} = n_{\mathrm{crit,hd}} \cdot \rho_{\mathrm{tr,hd}} \tag{8.1}$$

The number of critical scenarios is influenced by the human driver's (index hd, ego) driving behavior $B_{\mathrm{hd,ego}}$ as well as by the occurrence of surrounding circumstances that are not influenced by the driver (E_{te}, exposure of circumstances for potential hazards in the traffic environment):

$$n_{\mathrm{crit,hd}} = f_{\mathrm{crit}}\left(B_{\mathrm{hd,ego}}, E_{\mathrm{te}}\right) \tag{8.2}$$

The transition probability $\rho_{\mathrm{tr,hd}}$ is partially influenced by the skill $P_{\mathrm{hd,ego}}$ of the human driver as well as by the skills of other traffic participants (index tp):

$$\rho_{\mathrm{tr,hd}} = f_{\mathrm{tr}}\left(P_{\mathrm{hd,ego}}, P_{\mathrm{tp}}\right) \tag{8.3}$$

The link to the origin of critical scenarios as well as to the transition into an actual accident is mostly a multi-causal linkage of circumstances and can be described using a Swiss cheese model (Gründl 2005), as shown in Fig. 8.2. Every slice has multiple holes that can lead towards an adverse event, but accidents only occur when the holes line up and the appropriate adverse trigger occurs. In all other cases we are left with a near miss. Contrary to road traffic, critical scenarios in aviation and in health-care are well documented (Critical Incident Reporting System (CIRS)), and can be used to continuously improve safety. However, in road traffic, critical scenarios not resulting in accidents form a sort of "dark matter", as shown in Fig. 8.3. If this "dark matter" were known, one could determine far earlier to what extent automated driving can be involved in these critical scenarios. Furthermore, one could create a test benchmark from the transition probability of human drivers, defining the minimum controllability (or performance level) of an automated vehicle in critical situations. With regards to the Swiss cheese model, this would mean that the slices defining vehicle control could be individually determined.

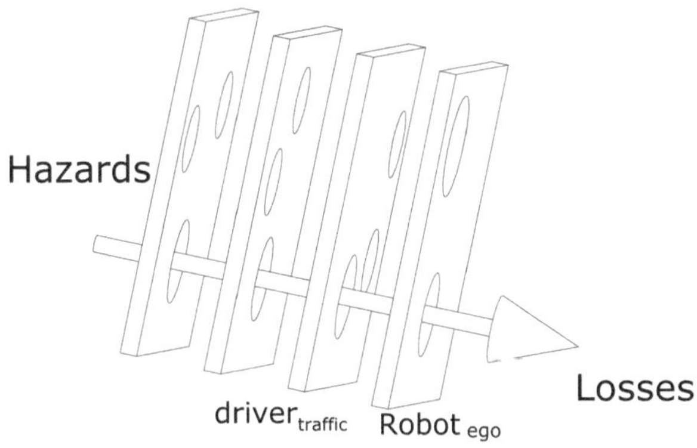

Fig. 8.2 The Swiss cheese model

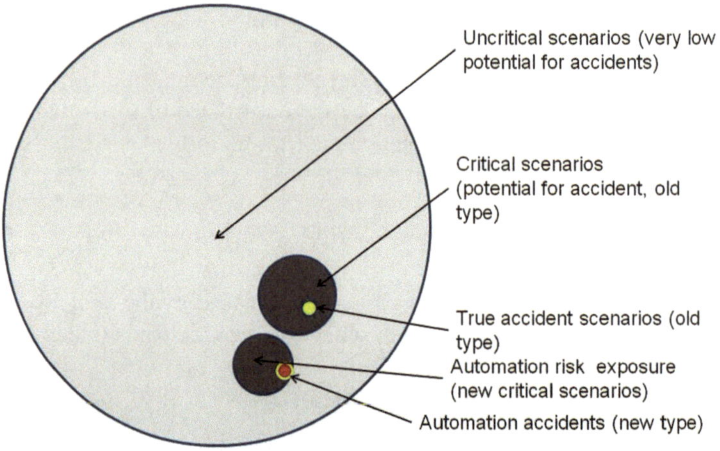

Fig. 8.3 The "dark matter" problem (Winner et al. 2016b)

This model used to forecast accidents could be adopted for automated driving (index *ad* instead of *hd*) if automated vehicles were to directly replace vehicles driven by humans and did not show any change in behavior. However, this is not realistic. On the one hand, autonomous cars are expected to be safer due to defensive driving compliant to the rules, already making a difference in the model probable. Here, most developers assume that the number of critical scenarios of the old type $n_{crit,ad,oT}$ (index oT, e.g. tailgating) will strongly decrease. On the other hand, the ground rules for driving will change fundamentally, which could reduce overall safety. Machine-perception is based on different principles, the behavior-generation does not comply with that of humans, and the behavior of other traffic participants will change under automated driving, as known from field tests conducted by Google (Urmson 2016). Therefore, critical scenarios of a new kind must be reckoned with, leading to accidents of a new kind.

This leads to the following coherences:

$$n_{\text{ac,hd}} = n_{\text{crit,ad,oT}} + n_{\text{crit,ad,nT}} \tag{8.4}$$

$$\begin{gathered} n_{\text{ac,hd}} = n_{\text{crit,ad,oT}} \cdot \rho_{\text{tr,ad,oT}};\, n_{\text{crit,ad,oT}} = f_{\text{crit}}\left(B_{\text{hd,ego}}, E_{\text{te,oT}}\right); \\ \rho_{\text{tr,ad,oT}} = f_{\text{tr}}\left(P_{\text{ad,ego,oT}}, P_{\text{te,oT}}\right) \end{gathered} \tag{8.5}$$

$$\begin{gathered} n_{\text{ac,ad,nT}} = n_{\text{crit,ad,nT}} \cdot \rho_{\text{tr,ad,nT}};\, n_{\text{crit,ad,nT}} = f_{\text{crit}}\left(B_{\text{ad,ego}}, E_{\text{te,nT}}\right); \\ \rho_{\text{tr,ad,nT}} = f_{\text{tr}}\left(P_{\text{ad,ego,nT}}, P_{\text{te,nT}}\right) \end{gathered} \tag{8.6}$$

In the Swiss cheese model, this means that the probability of passing through individual layers must be determined for autonomous driving. Under the "dark matter" model, the expectation is that the critical scenarios of the old type, the original dark matter, will lessen, and thus the overall accident rate should decrease, due to poor Swiss cheese transition probabilities. However, critical scenarios of the new type will be introduced with their own new risks and transition probabilities. These are the new risks caused by automation.

This means that it is not sufficient to only eliminate mistakes caused by human drivers, but that new mistakes caused by the introduction of automation must be considered as well. Thankfully, the equations can indicate a direction for the validation strategy:

The goal is to identify all relevant critical scenarios and then determine a transition probability or, in other words, the controllability

$$C = 1 - \rho_{\text{tr}} \tag{8.7}$$

for them. Obviously, the assumption of one critical scenario with only one transition probability does not apply to all types of accidents. Therefore, this approach must be extended to all types of accidents and scenarios. Presumably a transition probability must be modeled which depends on the criticality that is reached during a critical scenario, regarding the different categories of accident severity. To put it simply: the proof of safety is reduced to the proof that $n_{\text{ac,ad}} \leq n_{\text{ac,hd}}$. To solve this problem, the current dark matter must be "illuminated", meaning that preferably all critical scenarios must be found in order to determine to what extent automated driving is subjected to these scenarios and how well it is able to control them.

8.6 Derived Implementation Strategies

As it can be seen intuitively from the accident forecast model above, simplifying the driving task leads to a reduction of critical scenarios and to a higher controllability of these scenarios by the driving robot. The driving task can be simplified by

- Reducing the number of action alternatives,
- Reducing driving speed,
- Deployment only on simply-structured traffic areas such as Autobahn,
- Deployment only in either previously thoroughly test locations or in an assessable road system.

The testing effort can be significantly decreased by combining some of these measures for the system design. If the safety goals are to be actually achieved, the test cases and test results must first be proven in practice.

The currently dominating development paths are shown in Fig. 8.4. While most European vehicle manufacturers increase the level of automation in an incremental process and target the greatest possible extent of use, Google has directly started with fully autonomous driving, but limits itself to selected locations. In both approaches, the knowledge that is needed for the future universal deployment of automated vehicles is gained in an evolutionary process. The future will show which approach will be more successful, while the parallel evolution of both

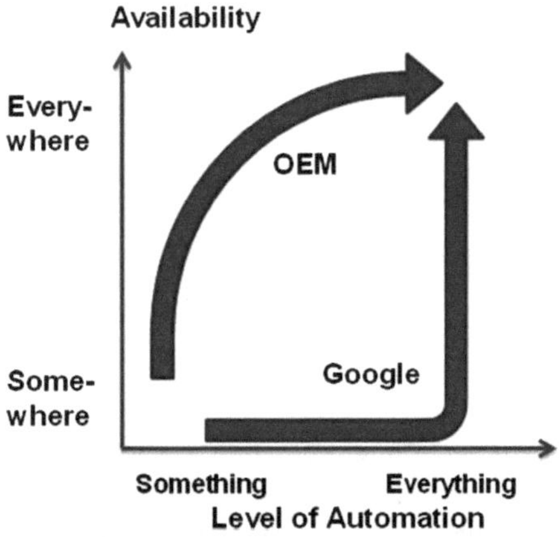

Fig. 8.4 Development paths to automated driving (Winner et al. 2016b)

approaches seems to be sensible, especially if different business and mobility models are pursued.

8.7 Potential Validation Concepts

As statistical validation prior to market launch is considered impossible, alternative validation concepts must be found. Possible approaches are cited from Wachenfeld and Winner (2016):

8.7.1 Reuse of Validated Functions

The first and easiest possibility for validating an automated vehicle's safety is to reuse known functions that already have been approved. Extended functionality must be separately validated; the less new functionality, the less effort. The incremental approach mentioned in the previous section provides very good fundamentals.

8.7.2 Accelerating the Validation Process

Even when pursuing an evolutionary approach on fully automating vehicles, new functionalities must still be validated. Principally, two setscrews exist in order to accelerate the validation process: Firstly, the "what", and secondly, the "how" can be changed. Which test cases are required and how are those tests conducted?

Approaches of Glauner et al. (2012) and Eckstein and Zlocki (2013) describe the identification process of relevant and critical situations in public traffic: During test drives or large-scale field tests, potentially critical situations are identified based on preassigned event classifications. These critical situations influence the generation of test situations, allowing for situations with low criticality to be neglected. This pooling of test cases is justified by the assumption that less critical situations are adequately represented by the critical ones. This currently leaves us with the unsolved task of finding a valid risk measure which allows us to rate situations in the first place and, secondly, to select critical situations. This approach precisely follows the approach of "illuminating" dark matter.

Schuldt et al. (2013) present us another approach for pooling test cases: They propose a generic generation of test cases using black-box-testing and combinatorics to cover the influencing factors on the system's safety as thoroughly as possible and to be as non-redundant and efficient as possible at the same time. This approach is based on statistical considerations without any knowledge or experience about the testing object. However, it still has the potential to reduce the amount of necessary test cases.

The approach described by Tatar and Mauss (2014) is suited for black-box-testing as well: test cases generation is formulated as an optimization problem. Thereby, the input parameters of a XiL-simulation are varied so as to optimize a rating function. Despite the challenge of creating a valid XiL-Simulation and the required rating function, this approach makes it possible to focus only on the test cases defined as relevant.

The use of formal methods (Mitsch et al. 2013) to deploy and test a safety concept represents a fourth theoretical approach. Just as with human-in-the-loop driving, a safety concept proven safe can make a complete test of a vehicle's functionality superfluous. Thus, a pooling of the test cases would be possible.

As an alternative to pooling test cases during test case generation, improving how tests are conducted can accelerate validation. Abstracting away from real-world driving and using different test tools always involves simplification, as shown in Fig. 8.5.

Figure 8.5 divides potential test tools into nine classes that differ in how they represent the vehicle or the environment. The driver is grouped with the vehicle, as he is seated in the vehicle and does not actively interfere with the automated drive.

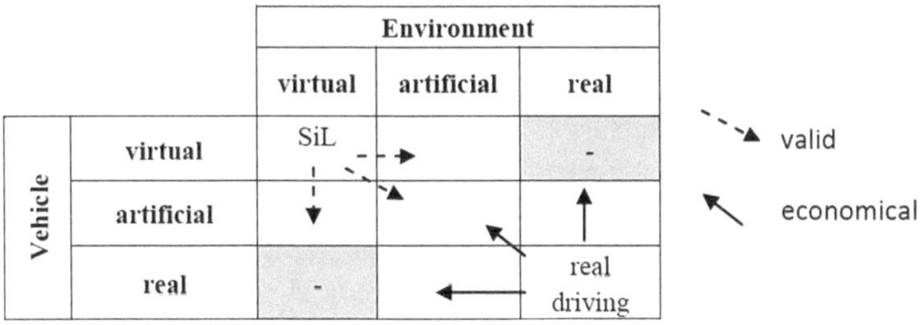

Fig. 8.5 Classification of test tools for testing automated vehicles (Wachenfeld and Winner 2015a)

Real-world test drives accurately represent the environment and the vehicle. Thus, real-world driving comes with the risk of accidents and their consequences. The environment is not controlled, so test situations originate from the coincidences of reality; therefore, the reproducibility of complex situations with other traffic participants is not possible. At the earliest, this test tool can be deployed with the first street-legal prototypes and thus this will not be used until the end of the development process.

An alternative is to test real vehicles in an artificial environment: This is equivalent to a ride on a test track, where the occurring "traffic situations" are artificial and the "traffic participants" are aware of the fact that they are part of a test. Reality is simplified in favor of safety, variability, observability, and reproducibility. Because of economic reasons, test cases undergo targeted testing and need not be driven by chance, as is the case for real-world test drives. However, creating the test environment requires additional time and money.

Furthermore, an artificial vehicle can drive in a real environment; the term artificial arises from the automated vehicle being equipped with a supervisor who has the ability to intervene into the driving task. This human-in-the-loop (HiL) supervisor can be a test driver with a steering wheel and pedals or a technical system that is superior to the automated system due to advanced (additional) sensors. If components are artificially depicted, the contact with reality is lessened but on the other hand, safety, reproducibility, and observability are improved.

Beside the ability to artificially design the vehicle and environment, tools exist that use a virtual representation of the real world in the form of a computer simulation. The two fields in Fig. 8.5 combining virtual and real systems are marked in gray, as they are technically not existent due to the fact that sensors and actuators have exactly the task to convert virtual to real signals and vice versa. A real radar sensor cannot sense a virtual environment and a virtual inverter cannot generate a real voltage.

Possible on the other hand is a combination of an artificial and a real environment or vehicle, for instance vehicle-in-the-loop (ViL) systems. Real components in a simulation are replaced with models, completing the circle of actions and reactions of environment and vehicle. Thereby, either the mentioned sensors or actuators are artificially stimulated (examples are simulation-based videos as stimulants for camera systems or dynamometers as stimulants for actuators), or the test tools directly simulate the power signals, an electromagnetic wave for instance, and try to depict real effects of sensors and actuators using models. See Bock (2012) or Hendriks et al. (2010) for further information. The meaningfulness of the test tools is questioned by use of models. To obtain valid propositions when using such models, one must prove that the models do not contain illegitimate simplifications; illegitimate is to be seen in a context of function and means that deviations from reality must be within the function's tolerance. If, however, this validity was proven, the test tool allows for greater safety during testing as parts of the environment and the vehicle would only meet in a virtual world. Due to these virtual components, the test tools are marked for greater validity, observability, and reproducibility. From an economical point of view, this test tool has the advantage that the virtual environment can be easily altered and updated to depict the vehicle in countless variations. An economic disadvantage can result from the validation of the models (see the following section). An advantage of this test tool is the ability to conduct tests based on the simulated vehicle in early development.

The last level of abstraction represents the combination of the virtual vehicle and the virtual environment: Here, the test tool referred to as software-in-the-loop (SiL) represents a closed control loop by modeling all relevant components in simulation. Contrary to the previous test tools, the entire test world is virtual. The tests are safe, more variable, observable, and reproducible. Furthermore, this tool can also be deployed in early stages of vehicle development. Hardware independence breaks the link to the real world and real time requirements, providing an additional economic advantage. Available computation power is the primary factor in time to test completion; simulations can be conducted day and night, as well as massively in parallel. Unfortunately, fully virtual tests suffer from increased abstraction from reality and require that every single model used within has been validated. Only if the validity of every model is proven virtual tests can be meaningful for validation. Thus, an economical consideration of simulation-based procedures must especially take into account the validation of the underlying models. This problem is especially affected by large knowledge gaps, as shown in the following section.

The same challenge exists in the use of formal methods. Regarding this matter, Mitsch et al. (2013) writes: "We (...) prove that collisions can never occur (as long as the robot system fits to the model)." This means that the reality of the models strongly influences the meaningfulness for formal methods, too. One special challenge and therefore a focus of current research is the formalization of sensor uncertainties and of traffic participant characteristics.

A discussion of the test tools shows the potential to accelerate the validation of fully automated vehicles: With the use of an artificially generated environment and vehicle, test cases can be specifically built up and tested. Furthermore, virtualization makes it possible to accelerate and parallelize tests limited only by the available computation power.

However, the discussion also shows that the validity and therefore the meaningfulness of the tests will become a challenge when introducing artificial and virtual components to automated vehicle testing.

8.8 The Challenge of Validity

Though methods exist that have the potential to effectively validate autonomous vehicles, these methods themselves must first be validated prior to any large-scale implementation. This requires proving the validity of the catalog of test scenarios and any models used for XiL-validation. The challenges for these are discussed separately.

8.8.1 Validity of the Test Catalog

A test catalog is only valid if its critical test cases are representative for future deployment of the system and if passing all test criteria are valid under future deployment conditions. One runs into the curse of dimensionality when trying to cover all critical cases. Even with parameterized abstractions one is quickly lost in the configuration space. There are roadway parameters (road geometry, roadway condition, roadside construction, the type and position of road signs, traffic signals and their condition, etc.), various weather conditions (solar elevation, rain and snow, temperature, range of vision, etc.), and an uncountable number of traffic participant constellations (variable inter-vehicular distances, speeds, alignments, intentions, behavior, and dynamic possibilities). Cautiously varying only the most important parameters for certain selected scenarios will still produce enough events to over-strain any test methodology, including software-in-the-loop tests. One must thus combine the influential parameters in some way. Monte-Carlo methods generate scenario parameters which capture the frequency of real-world occurrence. Here, the challenge is that the parameters are typically correlated, and exactly how must be established beforehand. This approach works in principle but can easily become too elaborate if a dense coverage is attempted and sparse test generation risk missing important scenarios.

A further approach decomposes tests according to the causes of failure. This approach combines the Swiss cheese model with a retrospective failure description of an accident according to Graab et al. (2008a). Graab presented five levels categorizing why an accident was not prevented (see Fig. 8.6). These can form a basis for test case decomposition and make it possible to only select relevant test cases for each level. The result is no longer a binary accident vs. non-accident, but rather an evaluation at every level, with the possibility that multiple levels co-exist

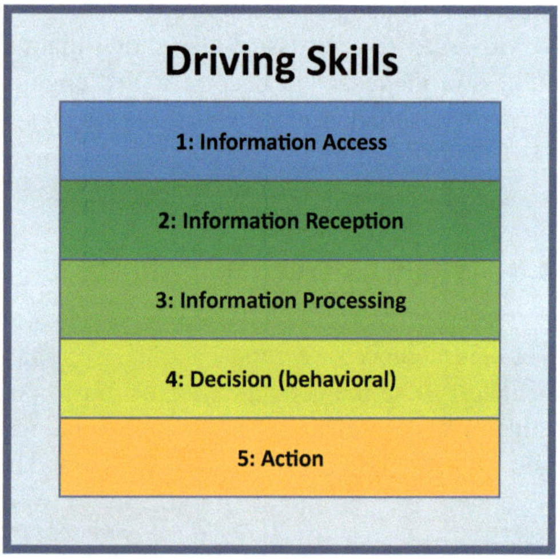

Fig. 8.6 Decomposition levels according to Graab et al. (2008b)

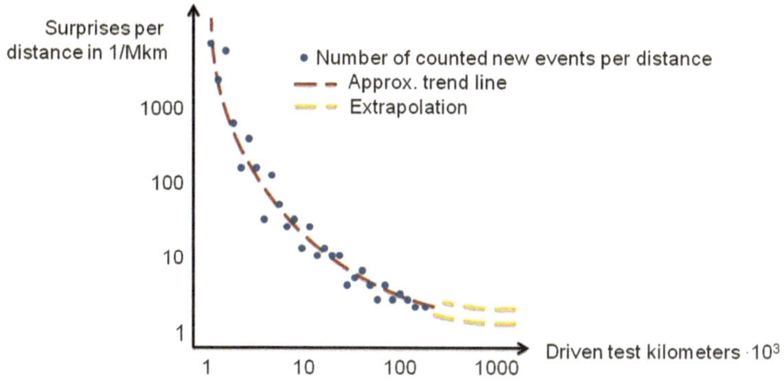

Fig. 8.7 Number of surprises per distance covered

in one test case. Furthermore, the passing criteria must be chosen so that criticality is no over- or under-emphasized. The authors would recommend to choose less critical fail criteria with further testing of the potential hazards in case of failing the test using representative scenarios.

Despite all theoretical approaches, there will always be doubts whether or not the test catalog is complete enough for safety certification. Real-world driving tests can be conducted in order to determine the maturity level of a test catalog. The frequency of surprises per testing distance is suitable as a measure of catalog maturity. An event is considered a surprise if the automated driver has reached an unwanted condition that is either outside of the specification or has not been included in the specification and is therefore not covered by the test catalog. This approach is similar to California's testing regulations, which require that all driver interventions must be published (State of California, Department of Motor Vehicles 2014). In most cases, interventions are due to missing test cases (or insufficient specifications) and are therefore surprises by definition, independent of whether they pose a threat. The inclusion of scenarios in the test catalog, either entirely or decomposed, leads to the steady improvement of the catalog and a decline in surprises per distance in subsequent drive tests. This is illustrated in Fig. 8.7 and is in line with Google reports. The maturity of the test catalog can be estimated from the trend of the progression. If certain critical assumptions are made, the remaining risk can estimated as well, where risk is calculated using the rate of surprises under previous tests. Besides drive tests against the targeted functionality, the VAAFO-concept described below can determine surprises from rides with previous functions and thus also improve the test catalog.

8.9 Validity of the Models

Beside the general conflict between cost and validity described above, we want to briefly cover the state-of-the-art in models and their validity. The simulation of vehicle dynamics can be used to attest model quality to a high degree, as vehicle dynamics are already validated during the homologation of ESC in various vehicles, for which the driving physics are correctly represented (Baake et al. 2014). The sensors that measure the vehicle dynamics are also simulated with sufficient quality. For environmental perception sensors, however, this correct representation with sufficient quality is missing. A challenge is posed by the need to simulate analog circuits of white-box models in real-time. Even if this problem was solved, the main challenge of simulating the environment relevant to the sensor, still remains. Even if modern raytracing algorithms reach a high degree of realism on graphics hardware (as is strikingly shown in modern game engines), they reach their limits when modeling environment sensors (Bernsteiner et al. 2015). Until now, translating the various optical effects present in a real camera image in a simulated camera model remains a difficult task. Accurately modeling radar sensors is even more difficult, as the received radar echo includes many multipath reflections from road surfaces, walls, and other vehicles which are superimposed according to phase. Generating correct raw signals requires that the environment is decomposed into surface elements the size of the signal wavelength, which is 4 mm for 77 GHz automotive radar. The surface normals and reflection coefficients would have to be specified for each element, and these parameters would have to be simultaneously accessible in memory, for every time-step in simulation. This shows that realistic radar simulations are presently impossible. As a first approach to solve this problem, Cao et al. (1999) proposes a gray-box simulation that contains the most important physical influences on the depiction of the environment onto the sensors. There is still a long way to go before environment sensors are considered to be validated for the purpose of official automotive validation, especially considering that validity metrics have not been defined.

A highly simplified model of environmental assessment may already suffice for behavioral modeling. Models for traffic participants are currently not particularly rich or adequate enough to convincingly represent scenarios. Furthermore, human driving behavior is expected to change when confronted with automated vehicles, at a minimum due to more cautious driving by the automated vehicles but also due to communication problems between human and automation will be relevant (Färber 2016). Determining how exactly human driving behavior will change prior to real-world testing seems impossible.

8.10 Acquiring Field Data

As is clear from previous discussion, a large knowledge gap exists that has to be filled with real-world driving data. Generally, the following methods for acquiring field data exist:

- Recording of a full test drive
  - With offline labeling
  - With automated labeling

- Recording of critical scenarios
  - With a trigger button controlled by the driver
  - With online automated labeling in the test vehicle

The last concept is part of the VAAFO (Virtual Assessment of Automation in Field Operation) concept (Wachenfeld and Winner 2015b).

Generally, it is irrelevant whether measurement data is acquired from active or emulated target functions, from previous (outdated) automated vehicles, or from manual drives, but its relevance declines in this order.

The first alternative to field data acquirement is a customer-oriented driving test, which has its limits when deployed with "real" customers. The financial effort for providing this technology and evaluating the results is already a challenge. The second concept is the technically most powerful but comes with privacy concerns.

VAAFO is an auto-labeling-tool that makes a retrospective comparison with real-world driving on the basis of a constantly restarted simulation. If significant discrepancies occur, a recording of values stored in a circular buffer is triggered. This can also be used to test new simulated functions that are not even possible with the present hardware or are not yet enabled because they are still under test. Thus, potential problems can be determined without the risk that they are appended to the test catalog, as previously discussed. Further details to this approach can be found in Wachenfeld and Winner (2015b).

8.11 Conclusion

Various methods that seem suitable for the validation of automated vehicles have been presented with the goal of validating automated driving without human supervision. Such methods are still far away from release, due to insufficiencies in both the "what" and the "how" of the validation process. The need to overcome both limitations has let to targeted research projects. The PEGASUS project[1] (01/2016–06/2019) is focused on the question "what", while the validation methods themselves ("how") are targeted by the EU ENABLE-S3 project[2] (05/2016–04/2019) under a joint ECSEL undertaking. The latter also addresses cyber-security, which has recently become a safety concern. The potential for cyber-crime and hacked automated vehicles is all too great to be neglected under the threat of modern terrorism.

Despite the best preparatory safety validation effort, real world deployment and validation during usage will show whether the reached safety will lead to acceptance by the exposed humans. However, conservative risk forecasts should allow

[1] http://www.pegasus-projekt.info/en/home

[2] http://www.enable-s3.eu/

for the appropriate scale of deployment to be made. If the predicted risk is below the yearly fluctuations of accident statistics and thus insignificant to risk placed on other traffic participants, then real-world travel distances can be accumulated that help improve the forecast until safety can be proven on a statistical basis. Horn and Watzenig (2016) model calculations, Wachenfeld (2016) shows that an introduction of automated vehicles compliant with this restriction will not decrease the speed of innovation, but rather can make very fast market penetration possible if the requirements of the expected safety and market demand are given.

References

Baake, U., Wüst, K., Maurer, M., Lutz, A.: Testing and simulation-based validation of ESP systems for vans. ATZ Worldwide. **116**(2), 30–35 (2014). doi:10.1007/s38311-014-0021-6

Baum, H., Kranz, T., Westerkamp, U.: Volkswirtschaftliche Kosten durch Straßenverkehrsunfälle in Deutschland. Berichte der Bundesanstalt für Straßenwesen Reihe M, Heft 208. Bundesanstalt für Straßenwesen, Bergisch Gladbach (2011)

Bernsteiner, D.-I.F.S., Magosi, Z., Lindvai-Soos, D.-I.D., et al.: Radar sensor model for the virtual development process. ATZelektronik Worldwide. **10**(2), 46–52 (2015)

Bock, T.: Bewertung von Fahrerassistenzsystemen mittels der vehicle in the loop-simulation. In: Winner, H., Hakuli, S., Wolf, G. (eds.) Handbuch Fahrerassistenzsysteme, pp. 76–83. Vieweg +Teubner Verlag, Wiesbaden (2012)

Cao, C.T., Kronenberg, K., Poljansek, M.: Adaptive transmission control. Google Patents. http://www.google.com/patents/US5954777 (1999)

Donges, E.: Aspekte der aktiven Sicherheit bei der Führung von Personenkraftwagen. AUTOMOB-IND. **27**(2), 183–190 (1982)

Eckstein, L., Zlocki, A.: Safety Potential of ADAS – Combined Methods for an Effective Evaluation ESV (2013)

Färber, B.: Communication and communication problems between autonomous vehicles and human drivers. In: Maurer, M., Gerdes, J.C., Lenz, B., Winner, H. (eds.) Autonomous Driving: Technical, Legal and Social Aspects, pp. 125–144. Springer, Berlin (2016)

Gasser, T.M., Arzt, C., Ayoubi, M., Bartels, A., Bürkle, L., Eier, J., Flemisch, F., Häcker, D., Hesse, T., Huber, W., Lotz, C., Maurer, M., Ruth-Schumacher, S., Schwarz, J., Vogt, W.: Rechtsfolgen zunehmender Fahrzeugautomatisierung: Gemeinsamer Schlussbericht der Projektgruppe. Berichte der Bundesanstalt für Strassenwesen - Fahrzeugtechnik (F), vol. 83. Wirtschaftsverl. NW Verl. für neue Wissenschaft, Bremerhaven (2012)

Glauner, P., Blumenstock, A., Haueis, A. (eds.): Effiziente Felderprobung von Fahrerassistenzsystemen UNI DAS e.V. (8. Workshop Fahrerassistenzsysteme) (2012)

Graab, B., Donner, E., Chiellino, U., Hoppe, M.: Analyse von Verkehrsunfällen hinsichtlich unterschiedlicher Fahrerpopulationen und daraus ableitbare Ergebnisse für die Entwicklung adaptiver Fahrerassistenzsysteme. Audi Accident Research Unit (AARU) (2008a)

Graab, B., Donner, E., Chiellino, U., Hoppe, M.: Analyse von Verkehrsunfällen hinsichtlich unterschiedlicher Fahrerpopulationen und daraus ableitbarer Ergebnisse für die Entwicklung adaptiver Fahrerassistenzsysteme. In: 3. Tagung Aktive Sicherheit durch Fahrerassistenz, 7.–8. April in Garching (2008b)

Gründl, M.: Fehler und Fehlverhalten als Ursache von Verkehrsunfällen und Konsequenzen für das Unfallvermeidungspotenzial und die Gestaltung von Fahrerassistenzsystemen (2005)

Hendriks, F., Tideman, M., Pelders, R., Bours, R., Liu, X.: Development tools for active safety systems: Prescan and VeHIL. In: Vehicular Electronics and Safety (ICVES), IEEE, QingDao (2010)

Horn, M., Watzenig, D. (eds.): Automated Driving: Safer and More Efficient Future Driving. Springer, Cham (2016)

Mitsch, S., Ghorbal, K., Platzer, A.: On provably safe obstacle avoidance for autonomous robotic ground vehicles. Proceedings of Robotics Science and Systems (RSS) (2013). Accessed 27 June 2014

NHTSA: Preliminary Statement of Policy Concerning Automated Vehicles. http://www.nhtsa.gov/staticfiles/rulemaking/pdf/Automated_Vehicles_Policy.pdf, access 06/2017 (2013)

Rasmussen, J.: Skills, rules, and knowledge; signals, signs, and symbols, and other distinctions in human performance models. IEEE Trans. Syst. Man Cybern. **SMC-13**(3), 257–266 (1983)

Reichart, G.: Menschliche Zuverlässigkeit beim Führen von Kraftfahrzeugen. VDI-Verlag, Düsseldorf (2001a)

Reichart, G.: Zuverlässigkeit beim Führen von Kraftfahrzeugen: Fortschrift-Berichte Nr. 7, VDI-Verlag. Diss., Technische Universität München (2001b)

SAE International Standard J3016: Taxonomy and Definitions for Terms Related to On-Road Motor Vehicle Automated Driving Systems. SAE International (2014)

Schnieder, E., Schnieder, L.: Verkehrssicherheit: Maße und Modelle, Methoden und Maßnahmen für den Straßen- und Schienenverkehr. VDI-Buch. Springer Vieweg, Berlin, Heidelberg (2013)

Schuldt, F., Saust, F., Lichte, B., Maurer, M., Scholz, S.: Effiziente systematische Testgenerierung für Fahrerassistenzsysteme in virtuellen Umgebungen. In: AAET (2013)

Schwarz, J.: Code of practice for the design and evaluation of ADAS PreVENT project. www.prevent-ip.org (2006)

Schwing, R.C., Albers, W.A.: Societal Risk Assessment: How Safe is Safe Enough? Springer, New York (2013)

State of California, Department of Motor Vehicles: Regulations for Testing of Autonomous Vehicles (2014)

statista.de: Sterbetafel: Deutschland, Jahre, Geschlecht, Vollendetes Alter. https://www-genesis.destatis.de/genesis/online/data;jsessionid=06C3C45E2F9F02CD74266ED33E17B88F.tom cat_GO_2_3?operation=abruftabelleAbrufen&selectionname=12621-0001&levelindex=1&levelid=1473423773217&index=1 (2013)

Statistisches Bundesamt: Verkehrsunfälle - Fachserie 8 Reihe 7 - 2013 (2013a)

Statistisches Bundesamt: Verkehr - Verkehrsunfälle - Fachserie 8 Reihe 7 2012, Wiesbaden (2013b)

Statistisches Bundesamt (Destatis): Verkehrsunfälle - Fachserie 8 Reihe 7 - 2015 (2015)

Tatar, M., Mauss, J.: Systematic Test and Validation of Complex Embedded Systems, Toulouse ERTS 2014 (2014)

Urmson, C.: Google Self-Driving Car Project. SXSW Interactive (2016)

Verband der Automobilindustrie: Automatisierung - Von Fahrerassistenzsystemen zum automatisierten Fahren. https://www.vda.de/de/services/Publikationen/automatisierung.html (2015). Accessed 8 Sept 2016

Vorndran, I.: Unfallstatistik-Verkehrsmittel im Risikovergleich Wirtschaft und Statistik (12) (2010)

Wachenfeld, W.: How stochastic can help to introduce automated driving. Dissertation, Technische Universität Darmstadt (2016)

Wachenfeld, W., Winner, H.: Die Freigabe des autonomen Fahrens. Freigabe des autonomen Fahrens. In: Maurer, M., Gerdes, J.C., Lenz, B., Winner, H. (eds.) Autonomes Fahren, pp. 439–464. Springer, Berlin (2015a)

Wachenfeld, W., Winner, H.: Virtual assessment of automation in field operation: a new runtime validation method. In: UNI DAS e.V (ed.) 10. Workshop Fahrerassistenzsysteme. ISBN 978-3-00-050746-5, http://www.uni-das.de/images/pdf/veroeffentlichungen/abs-03-wachenfeld.pdf, access 06/2017 (2015b)

Wachenfeld, W., Winner, H.: The release of autonomous vehicles. In: Maurer, M., Gerdes, J.C., Lenz, B., Winner, H. (eds.) Autonomous Driving: Technical, Legal and Social Aspects, pp. 425–449. Springer, Berlin (2016)

Winner, H., Hakuli, S., Lotz, F., Singer, C. (eds.): Handbuch Fahrerassistenzsysteme: Grundlagen, Komponenten und Systeme für aktive Sicherheit und Komfort. Springer Fachmedien Wiesbaden, Wiesbaden (2015)
Winner, H., Hakuli, S., Lotz, F., Singer, C. (eds.): Handbook of Driver Assistance Systems: Basic Information, Components and Systems for Active Safety and Comfort. Springer International Publishing, Cham (2016a)
Winner, H., Wachenfeld, W., Junietz, P.: (How) Can safety of automated driving be validated? In: 7. Grazer Symposium Virtuelles Fahrzeug, Graz (2016b)